Johann Baumeister

Stable Solution of Inverse Problems

Advanced Lectures in Mathematics

Edited by Gerd Fischer

Jochen Werner
Optimization. Theory and Applications

Manfred Denker
Asymptotic Distribution Theory
in Nonparametric Statistics

Klaus Lamotke
Regular Solids and
Isolated Singularities

Francesco Guaraldo, Patrizia Macrì,
Alessandro Tancredi
Topics on Real Analytic Spaces

Ernst Kunz
Kähler Differentials

Johann Baumeister
Stable Solution of Inverse Problems

Johann Baumeister

Stable Solution of Inverse Problems

Friedr. Vieweg & Sohn   Braunschweig/Wiesbaden

AMS Subject Classification: 35 R 25, 35 R 30, 45 A 05, 45 L 05, 65 F 20

1987
All rights reserved
© Friedr. Vieweg & Sohn Verlagsgesellschaft mbH, Braunschweig 1987

Produced by Lengericher Handelsdruckerei, Lengerich

ISBN-13: 978-3-528-08961-0   e-ISBN-13: 978-3-322-83967-1
DOI: 10.1007/ 978-3-322-83967-1

PREFACE

These notes are intended to describe the basic concepts of solving inverse problems in a stable way. Since almost all inverse problems are ill-posed in its original formulation the discussion of methods to overcome difficulties which result from this fact is the main subject of this book.

Over the past fifteen years, the number of publications on inverse problems has grown rapidly. Therefore, these notes can be neither a comprehensive introduction nor a complete monograph on the topics considered; it is designed to provide the main ideas and methods. Throughout, we have not striven for the most general statement, but the clearest one which would cover the most situations.

The presentation is intended to be accessible to students whose mathematical background includes basic courses in advanced calculus, linear algebra and functional analysis. Each chapter contains bibliographical comments. At the end of Chapter 1 references are given which refer to topics which are not studied in this book.

I am very grateful to Mrs. B. Brodt for typing and to W. Scondo and U. Schuch for inspecting the manuscript.

Frankfurt/Main, November 1986   Johann Baumeister

TABLE OF CONTENTS

VIII

PART I

BASIC CONCEPTS

In the first part we shall show that inverse problems appear in various fields of applied sciences and that the mathematical formulation of these problems is much the same for various applications. The main aspects, characteristics and solution concepts of inverse problems are discussed.

Chapter 1: Introduction

In this section we make a classification of the types of problems which arise in mathematical modelling of natural processes, consider several examples of inverse problems in different fields of applied sciences and discuss general properties of these problems.

1.1 Inverse problems

Suppose that we have a mathematical model of a physical process. We assume that this model gives a description of the system behind the process and its operating conditions and explains the principal quantities of the model:

input, system parameters, output

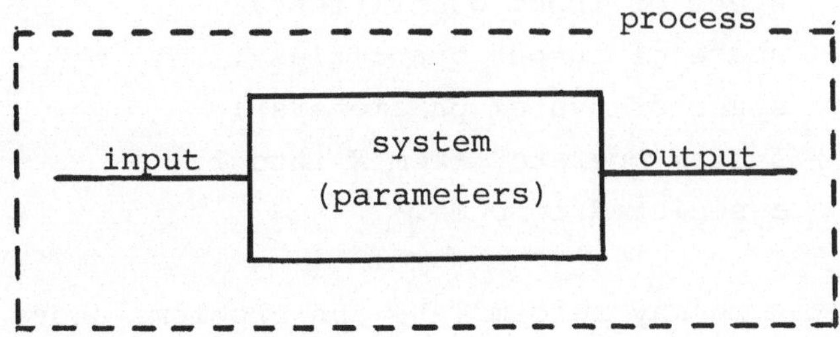

Figure 1

In the most cases the description of the system is given in terms of a set of equations (ordinary and/or partial differential equations, integral equations,...), containing certain parameters.

The analysis of the given physical process via the mathematical model may be separated into three distinct types of problems.

(A) The direct problem. Given the input and the system parameters, find out the output of the model.

(B) The reconstruction problem. Given the system parameters and the output, find out which input has led to this output.

(C) The identification problem. Given the input and the output, determine the system parameters which are in agreement with the relation between input and output.

We call a problem of type (A) a direct (or forward) problem since it is oriented a cause-effect sequence. In this sense problems of type (B) and (C) are called inverse problems because they are problems of finding out unknown causes of known consequences. It is immediately clear that the solution of one of the problems above involves a treatment also of the other problems. A complete discussion of the model by solving the inverse problems is the main objective of inverse modelling.

We give a mathematical description of the input, the output and the system in functionalanalytic terms.

$$\begin{aligned}
&X \quad \text{space of input quantities ;}\\
&Y \quad \text{space of output quantities ;}\\
&\mathbb{P} \quad \text{space of system parameters ;}\\
&A(p) \ \text{system operator from } X \text{ into } Y\\
&\qquad \text{associated to } p \in \mathbb{P} .
\end{aligned}$$

In these terms we may reformulate the problems above in the following way:

(A) Given $x \in X$ and $p \in \mathbb{P}$, find $y := A(p)x$.

(B) Given $y \in Y$ and $p \in \mathbb{P}$, solve the equation

$$Ax = y \qquad (x \in X)$$

where $A := A(p)$.

(C) Given $y \in Y$ and $x \in X$, find $p \in \mathbb{P}$ such that

$$A(p)x = y.$$

At first glance, the direct problem seems to be solved much more easier than the inverse problems. However, for the computation of $y := A(p)x$ it may be necessary to solve a differential or integral equation, a task which may be of the same complexity as the solution of the equations in the inverse problems.

In certain simple examples inverse problems can be converted formally into a direct problem. For example, if A has a known inverse then the reconstruction problem is solved by $x := A^{-1}y$. However, the explicit determination of the inverse does not help if the output y is not in the domain of definition of A^{-1}. This situation is typical in applications due to the fact that the output may be only imprecisely known and/or distorted by noise.

The reconstruction problem is the main objective of these notes. The linear case, that is if A is a linear map, has been studied extensively and its theory is well-developed. We shall give a detailed presentation of this theory. The situation in the nonlinear case is somewhat less satisfactory. Linearization is very successful to find an acceptable solution to a nonlinear problem but in general this principle provides only a partial answer.
The identification problem in general setting is rather difficult since it is in almost all cases a highly nonlinear problem. We shall give only a brief introduction to this field.

1.2 Some examples of inverse problems

We present several examples of inverse problems in different fields of applied science to find out the characteristic questions and problems in studying inverse problems. Emphasis is given to the way inverse problems appear in science rather than to rigorous mathematical formulations and proofs.

(E1) Reconstruction of unknown forces

Dynamic forces acting on a mechanical system cannot always be measured directly for example by installing some kind of dynamometer. Hence there is a great temptation to pose the following problem:

> Find the unknown dynamic force having measured
> vibration response of a system, parameters of
> which are considered to be known.

Applications are given for various mechanical systems (ball mills, off-shore platforms).

If we consider a one-degree-of freedom mechanical system the behavior of such a system may be described by the following ordinary differential equation

$$(1.1) \qquad m\ddot{y} + ky = x(t) \quad , \quad t > 0 \ ,$$

where m is the mass of the system, k is the stiffness constant, dots indicate derivatives, y is a function describing the displacement and x is the function of unknown forces. If the displacement function y is known precisely then the reconstruction problem may be solved by plugging in y into the differential equation (1.1); this gives

$$(1.2) \qquad x(t) = m\ddot{y}(t) + ky(t) \quad , \quad t > 0 .$$

But if y is contaminated by some noise η the same procedure cannot be applied if η is not twice differentiable as it is typical in applications. Even if η would be regular enough $m\ddot{\eta}$ might be a highly oscillating function.

Example 1.1

Consider the model (1.1) with $m = k = 1$, displacement function $y(t) := e^{-t}$. If we reconstruct the (unknown) force x from the perturbed data

$$\tilde{y}(t) := e^{-t} + a \sin wt \quad , \quad t > 0,$$

by the formula (1.2) we obtain as a result

$$\tilde{x}(t) = 2e^{-t} + a(1-w^2) \sin wt \quad , \quad t > 0.$$

We see that $\tilde{x}$ is a bad reconstruction if w is large. *

(E2) Reconstruction of signals

If we consider the distorsion of a true signal x by a physical device the observed signal y is given in a simple time-invariant model by

$$y(t) = \int_{-\infty}^{\infty} h(t-s)x(s)ds \quad , \quad t \in \mathbb{R},$$

where h is the so called apparatus function. The reconstruction problem consists in solving the equation

(1.3)
$$Ax = y$$
$$\text{where } (Ax)(t) := h * x(t) := \int_{-\infty}^{\infty} h(t-s)x(s)ds, \ t \in \mathbb{R}.$$

An equation of the special form (1.3) is called a convolution equation and the process of solving (1.3) is called deconvolution. Such equations appear in various fields, let us mention a few examples: image processing, optical information processing, image restoration, deconvolution of seismic data.

Signals $f : \mathbb{R} \longrightarrow \mathbb{R}$ which are square-integrable posess a Fouriertransform $\hat{f}$, defined by

$$\hat{f}(w) := \frac{1}{\sqrt{2\pi}} \int_{-\infty}^{\infty} f(t)e^{-iwt}dt, \quad w \in \mathbb{R} ,$$

in a little bit inexact way. $\hat{f}$ is known as the amplitude spectrum of f and is itself square-integrable. The inversion formula

$$f(t) = \frac{1}{\sqrt{2\pi}} \int_{-\infty}^{\infty} \hat{f}(w) e^{iwt} dw, \quad t \in \mathbb{R},$$

allows to consider signals in time space (f, t "time") and frequency space ($\hat{f}$, w "frequency") without loosing information. Using the convolution theorem for Fourier transforms we obtain from (1.3) the simple equation

$$(1.4) \qquad \frac{1}{\sqrt{2\pi}} \hat{\hat{h}}\hat{x} = \hat{y}.$$

Since $\lim_{|w| \to \infty} \hat{h}(w) = 0$ due to the fact that $\hat{h}$ is square-integrable there exist signals in the image space of the convolution map A such that their difference is arbitrary small, even if they correspond to widely different objects. Such disparate solutions, therefore, become indistinguishable in the presence of noise. An interesting special case is given if h is the lowpass filter

$$\hat{h}(w) = \begin{cases} 1, & |w| \leq \Omega \\ 0, & |w| > \Omega \end{cases}$$

where Ω is the highest frequency transmitted by the device.

(E3) Image reconstruction from projections

Let f describe the density of a medium in a region $\Omega \subset \mathbb{R}^2$. The combined effects of scattering and absorption result in a exponential attenuation of a beam of x-ray photons as it passes through the medium. If

I_o is the input intensity of the beam of x-ray photons the output intensity of the beam is given by

$$I_o \exp (- \int_L f(z) dz)$$

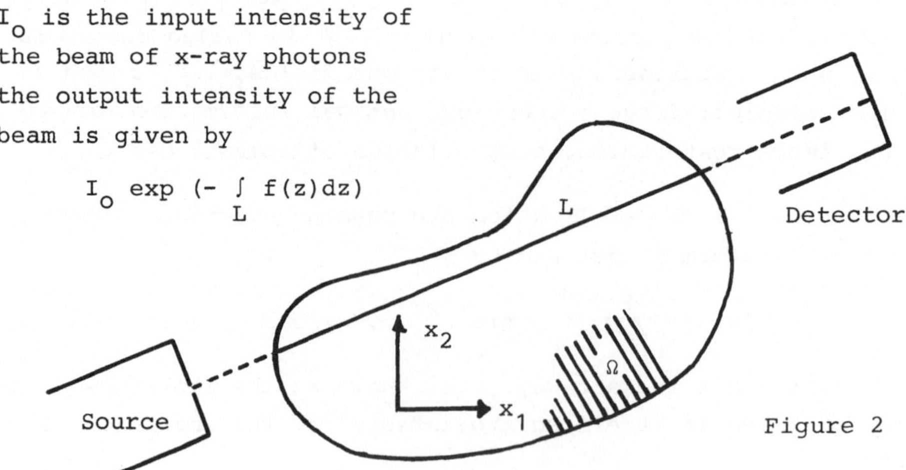

where L is the beam path. The line integral

$$\int_L f(z)\,dz$$

is called a projection. By moving the source of the beam and the detector around the medium it is possible to obtain a set of projections. Then an appropriate inversion algorithm is applied to recover an approximation to the density distribution f. By stacking several transverse sections of a body, the two-dimensional information may be converted to a three-dimensional information.

This method has been applied in various fields of applications:

- x-ray tomography in medicine (determination of tumors) ;

- geophysical tomography (determination of subsurface structure) ;

- optical tomography (flow field diagnostics).

Let us give a more mathematical formulation of the problem. A beam path L may be parametrized by an angle and a distance in the following way:

Figure 3

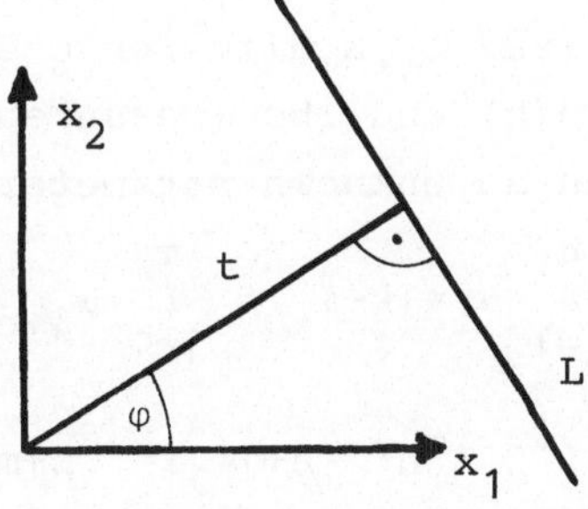

$$L = L_{t,\varphi} = \{z \in \mathbb{R}^2 \mid z = tu(\varphi) + su^{\perp}(\varphi),\ s \in \mathbb{R}\}$$

where $t \in [0,\infty)$, $\varphi \in (-\frac{\pi}{2}, \frac{\pi}{2})$, $u(\varphi) = (\cos\varphi, \sin\varphi)$,

$$u^{\perp}(\varphi) = (-\sin\varphi, \cos\varphi).$$

Then the line integral corresponding to $L_{t,\varphi}$ may be written as follows:

8

$$(1.5) \qquad R_\varphi f(t) := \int_{-\infty}^{\infty} f(tu(\varphi) + s\, u^\perp(\varphi))\,ds.$$

We refer to $R_\varphi f$ as the radiograph of f in the direction perpendicular to $u(\varphi)$. The function Rf, defined by

$$(1.6) \qquad Rf(\varphi,t) := R_\varphi f(t) \quad, \quad t \in [0,\infty), \quad \varphi \in (-\tfrac{\pi}{2}, \tfrac{\pi}{2})$$

is called the Radon transform of f. Thus, the solution of the reconstruction problem by projections consists in an inversion of the operator R, defined on a suitable space of functions. It turns out - theoretically speaking - that a finite number of x-rays taken from a finite number of different angles are never sufficient to provide the required information; on the other hand, inversion methods work rather well in practice if the projections are known for regular spaced values $\varphi \in (-\tfrac{\pi}{2}, \tfrac{\pi}{2})$ (complete tomography). The situation is much more difficult when the angles are restricted to lie in an interval $(-\theta,\theta)$ where $\theta < \frac{\pi}{2}$ (limited angle tomography).

(E4) System identification

The problem of identifying an "unknown" system among a class of systems is very important. Here we consider a system which is characterized by a difference equation relating the input time series u(k) and the (measured) process output time series v(k) through an unknown parameter vector p:

$$(1.7) \quad \left\{ \begin{array}{l} v(k) + \displaystyle\sum_{i=1}^{n} \alpha_i v(k-i) = \sum_{j=0}^{m} \alpha_{j+n+1}\, u(k-d-j), \ k \in \mathbb{N} \ , \\[2ex] p = (\alpha_1,\ldots,\alpha_n,\ \alpha_{n+1},\ldots,\alpha_{n+m+1})^t \end{array} \right.$$

Here and in the following "t" denotes transposition of vectors and matrices.
The equation (1.7) can be written as

$$a_k^t p = y_k \quad, \qquad k \in \mathbb{N},$$

$$(1.8)$$

$$\text{where } a_k^t := (-v(k-1),\ldots,-v(k-n),u(k-d),\ldots,u(k-d-m))$$

$$y_k := v(k) \ , \ k \in \mathbb{N}.$$

If we consider a large but finite segment of the time series the problem consists in solving a large and overdetermined system of equations. Since the right hand side y_k is a component of the (measured) process output the system is inconsistent. If p_k is any approximation of the parameter vector p the error in the (k+1)-th equation is

$$e_{k+1} = a_{k+1}^t \, p_k - y_{k+1}$$

We shall see in Section 8.5 that the update rule

$$(1.9) \qquad p_{k+1} := p_k - \lambda_{k+1} \, e_{k+1} \, a_{k+1}$$

leads to an algorithm which has some nice properties if the parameter λ_{k+1} is appropriate chosen.

Example 1.2

Consider the following special situation:
n = 0, m = 2, d = 1, k = 1,...,8.
The input sequence is a (random) sequence of zeros and ones.
Input and output data are summarized in the following table:

u(l)	0 0 0 1 1 0 0 1 1 1 0
v(k)	2 6 5 1 2 6 7 5

Table 1

The uniquely determined solution vector p^t is given by (2,4,1). The iteration (1.9) with initial guess $p_o = 0$ and step size rule $\lambda_k = 1$ gives the following approximation p_8 for p:

$$p_8^t = (2.25,\ 3.875,\ 1.125).$$

Using this vector p_8 as a new starting vector p_o we obtain the following approximation p_{16} for p:

$$p_{16}^t = (2.04688,\ 3.97656,\ 1.02344).$$

If we continue in this way we arrive at

$$p_{59}^t = (2.00000,\ 4.00000,\ 1.00000) \qquad *$$

(E5) Compartment analysis: an inverse problem

A compartment is an amount of some material which acts kinetically like a distinct, homogeneous, well-mixed amount of material. A compartmental system consists of one or more compartments which interact by exchanging material. There may be inputs into one or more compartments from outside the system and there may be excretions from the compartments of the system.

If we present the i-th compartment by a box characterized by the size q_i , the transfer constants λ_{ij} and the input v_i and if we diagramm the compartmental system by directed lines the system may be drawn as follows:

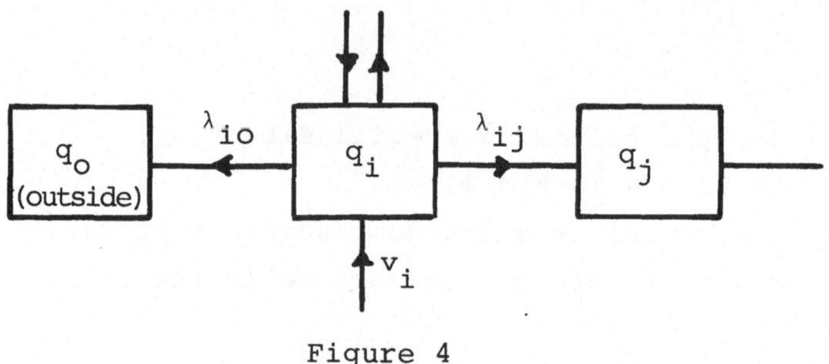

Figure 4

It is possible to express the compartment equations describing the flux of material in the system by a system of differential equations:

$$(1.10) \quad \begin{cases} q' = Mq + Bv(t) \quad , \quad t > 0 , \\ q(o) = \theta \end{cases}$$

where $M = (m_{ij}) \in \mathbb{R}^{n,n}$ (n compartments) with

$$m_{ij} = \begin{cases} \lambda_{ij} & , \text{ if } i = j \\ -(\lambda_{io} + \sum\limits_{j \neq i} \lambda_{ij}) & , \text{ if } i \neq j \end{cases} \quad ;$$

B is an n × 1 matrix which shows the possibilities of injection into the system.

The description of the system is completed by an observation equation

$$(1.11) \qquad y(t) = Cq(t) \quad , \ t > o,$$

where C is a $k \times n$ matrix, $k \leq n$.

The inverse problem of compartment theory may be posed as follows:

$$(1.12) \qquad \begin{array}{l} \text{Given the observation } y, \text{ determine each} \\ \text{coefficient } \ m_{ij} \text{ of the matrix } M. \end{array}$$

The analysis of this problem can be divided into two stages:

i) Check if all non-zero coefficients m_{ij} can be

uniquely determined by the observation (identifiability).

ii) Design an algorithm to compute the coefficients.

The input-output behavior of the compartment system represented by (1.10) and (1.11) is completely described by

$$(1.13) \qquad y(t) = \int_{o}^{t} Ce^{M(t-s)} Bv(s)ds \quad , \quad t > o.$$

Let f be a function on $[0,\infty)$; the Laplace transform $\check{f}$ of f is given by

$$\check{f}(s) = \int_{o}^{\infty} e^{-st} f(t)dt \quad , \quad s > o,$$

whenever this integral converges. Using the convolution theorem for Laplace transformations we obtain from equation (1.13)

$$(1.14) \qquad \check{y}(s) = C(sI-M)^{-1}B \quad , \ s > o.$$

The function

$$T: \ [0,\infty) \ni s \longrightarrow C(sI-M)^{-1}B \in {\rm I\!R}^{k,l}$$

is called the transfer function matrix. The investigation how many coefficients of M are determined by this function leads to identifiability results.

Example 1.3

n = 2, l = 1, k = 1

Figure 5

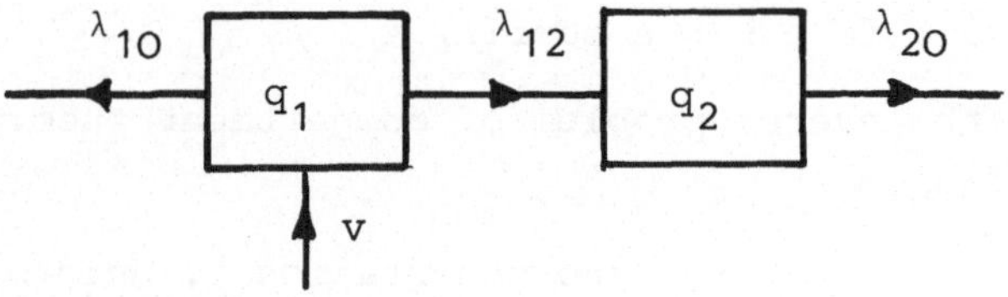

The associated system is given by

$$\begin{pmatrix} q_1' \\ q_2' \end{pmatrix} = \begin{pmatrix} -(\lambda_{10} + \lambda_{12}) & 0 \\ \lambda_{12} & -\lambda_{20} \end{pmatrix} \begin{pmatrix} q_1 \\ q_2 \end{pmatrix} + \begin{pmatrix} 1 \\ 0 \end{pmatrix} v(t)$$

$$y(t) = (1 \; 0) \begin{pmatrix} q_1(t) \\ q_2(t) \end{pmatrix} = q_1(t) \quad , \quad t > o.$$

Since the transfer function of the system is given by

$$T(s) := \frac{1}{s + \lambda_{10} + \lambda_{12}}$$

we see that only the sum $\lambda_{10} + \lambda_{12}$ can be determined by T.   *

Compartment theory has been successful applied to tracer experiments in biology and biochemistry.

(E6) <u>Identification of a diffusion coefficient</u>

The inverse problem in groundwater flow consists in finding out from a known steady state piezometric head the transmissivity of a porous medium. A two-dimensional model for the flow in a region $\Omega \subset \mathbb{R}^2$ is governed by the diffusion equation

(1.15) $- \operatorname{div}(a \, \nabla u) = f$

where u is the piezometric head, a is the (unknown) transmissivity and f is a given function. The inverse problem may be stated as follows:

(1.16)   Given the state u, find the coefficient a
which is in agreement with equation (1.15).

If we assume sufficient regularity the state equation (1.15) may be formulated as

(1.17) $- \nabla a \cdot \nabla u - a\Delta u = f$

and we see that the inverse problem consists in solving the first-order (hyperbolic) equation (1.17). The basic objective is therefore to derive conditions under which the hyperbolic problem is guaranteed to have a unique solution and the characterization of its dependence on the relevant parameters of the problem. It is well-known that the characteristics of the system ("curves of steepest ascent in u") play a fundamental role in the solution of the equation (1.17). For uniqueness of the parameter a one can show that a must be known along a line Γ_1 crossed by all streamlines (u = const). This line Γ_1 can be a portion of the boundary Γ of the region Ω. Then Γ_1 may be considered as inflow region of the flow (starting region for characteristics).

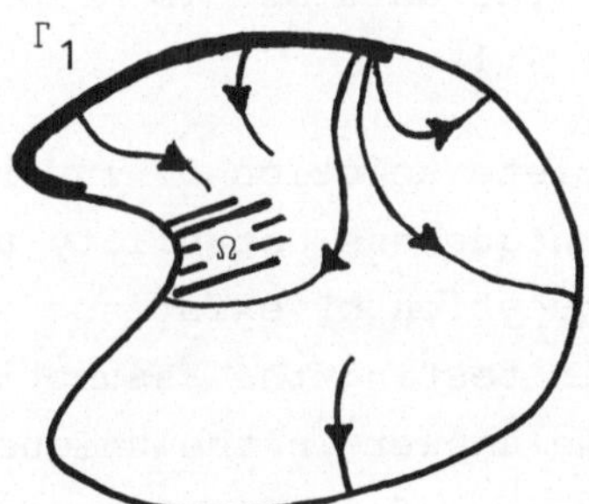

Figure 6

Example 1.4

If we consider the one-dimensional case only the equation (1.17) looks as follows:

$$- a'u' - au'' = f(x) \quad , \quad x \in \Omega \subset \mathbb{R}^1 .$$

A formal solution is given by

$$(*) \quad a(x) = \frac{1}{u'(x)} \{a(\bar{x})u'(\bar{x}) + \int_{\bar{x}}^{x} f(s)ds\} \quad , \quad x \in \Omega.$$

Therefore, if u'(x) is bounded away from zero over Ω and an "initial" value of a at $\bar{x} \in \bar{\Omega}$ is given, it is obvious that a unique solution for the inverse problem exists.

However, if u' vanishes at some points in $\bar{\Omega}$, we see by discussing the formula (*) that we may have nonuniqueness and/or nonexistence in the inverse problem.   *

1.3 Analysis of inverse problems

Inverse modelling involves the estimation of the solution of an equation from a set of observed data. The theory falls into two distinct parts. One deals with the ideal case in which the data are supposed to be known exactly and complete (perfect data). The other treats the practical problems that are created by incomplete and imprecise data (imperfect data). It might be thought that an exact solution to an inverse problem with perfect data would prove also useful for the practical case. But it turns out in inverse problems that the solution obtained by the analytic formula is very sensitive to the way in which the data set is completed and to errors in it (see Example 1.1).

In a complete solution of inverse problems the questions of existence, uniqueness, stability and construction are to consider. The question of existence and uniqueness is of great importance in testing the assumption behind any mathematical model. If the answer in the uniqueness question is no, then we know that even perfect data do not contain enough information to recover the physical quantity to be estimated. In the question of stability we have to decide if the solution depends continuously on the data. Stability is necessary if we want to be sure that a variation of the given data in a sufficiently small range leads to an arbitrary small change in the solution. This concept was introduced by HADAMARD in 1902 in connection with the study of boundary value problems and he designated unstable problems ill-posed.Almost all inverse problems of interest are illposed in this sense. When solving ill-posed problems numerically we must certainly expect some difficulties, since any errors act as a perturbation on the original equa-

tion and so may cause arbitrarily large variations in the solution. The same effect have observational errors. Since errors cannot be completely avoided there may be a range of plau-
sible solutions and we have to find out a reasonable solution.
These ambiguities in the solution of inverse problems which
are unstable can be reduced by incorporating some sort of a-
priori information that limits the class of allowable solu-
tions.By a priori-information we mean an information which has
been obtained independently of the observed values of the data.
This a priori-information may be given as a deterministic or a
statistical information. We shall restrict ourselves to deterministic considerations.

<u>Bibliographic comments</u>

Inverse problems have received a great deal of interest in
different fields of applications during recent years - see [4]
[27], [51], [53], [56], [90] . The formulation of inverse problems date
back to Lord RAYLEIGH (1877), HERGLOTZ (1914), LANGER (1933),
and others. Recent investigations have been undertaken in the
following areas: Reconstruction of unknown sources [83] and signals [7], image reconstruction [18], computer tomography in medicine [95], geophysics [81], and optics [92], compartment analysis
[1], identification of parameters in static [88] and dynamic processes [10], seismic exploration [14], electrocardiography [20]
and magnetocardiography [105], inverse scattering [21], [19], radiography [58], evolution backwards in time [16], inverse heat conduction [63], inverse melting problems [63]. An extensive bibliog-
raphy on inverse and ill-posed problems is contained in [56].

Chapter 2: Ill-posed problems

This section is devoted to a preliminary discussion of the stability problem. We shall give a definition of ill-posed problems and sketch the main idea to restore stability in ill-

2.1 General properties

Inverse problems in the sense of Section 1.1 involve operator equations. As a model equation we consider the following equation

$$(2.1) \qquad Gu = w \qquad (u \in U, \ w \in W)$$

where G is a mapping from the (solution-)space U into the (data-)space W; U and W are assumed to be subsets of normed spaces X and Y respectively.

Definition 2.1

The problem to solve the equation (2.1) is <u>well-posed</u> (in the sense of HADAMARD) if the inverse map $G^{-1} : W \longrightarrow U$ exists and is continuous, otherwise the problem is called <u>ill-posed</u>.

The classical example for an ill-posed problem which was given by HADAMARD is an initial value problem for the Laplace equation:

Example 2.2

Find a function $u = u(x,y)$ satisfying

$$(2.2) \qquad \frac{\partial^2 u}{\partial x^2} + \frac{\partial^2 u}{\partial y^2} = 0 \qquad , \quad \text{in } \mathbb{R} \times (0,\infty)$$

$$u(x,o) = 0 \quad , \quad \frac{\partial u}{\partial y}(x,o) = w(x), \quad x \in \mathbb{R}$$

If we choose as data

$$w(x) := w_n(x) := n^{-1} \sin(nx) \ , \quad x \in \mathbb{R} \ , \quad n \in \mathbb{N},$$

and

$$w(x) := w_o(x) := 0 \qquad , \ x \in \mathbb{R} \ ,$$

we obtain as solutions

$$u_n(x,y) = n^{-2} \sin(nx) \ \sinh(ny), \ (x,y) \in \mathbb{R} \times (0,\infty), \ n \in \mathbb{N},$$

and

$$u_o(x,y) = 0 \qquad , \ (x,y) \in \mathbb{R} \times (0,\infty),$$

respectively; the solutions are determined uniquely by the data. We see immediately:

i)   The sequence $(w_n)_{n \in \mathbb{N}}$ converges uniformly to w_o.

ii) $\overline{\lim\limits_n} \ |u_n(x,y) - u_o(x,y)| = \infty$ for all $(x,y) \in \mathbb{R} \times (0,\infty), x \neq 0$.

iii) The sequence $(u_n)_{n \in \mathbb{N}}$ doesn't converge to u_o in any reasonable topology, that is, the solution u of (2.2) doesn't depend continuously on the data w in any reasonable topology.   *

According to definition 2.1 we may distinguish between three types of non-well-posedness. If w is not in the range of the map G, then (2.1) has no solution (<u>nonsolvability</u>). If G is not one-to-one, G^{-1} doesn't exist and the equation (2.1) may have several solutions if $w \in G(U)$ (<u>ambiguity</u>). Finally, if G^{-1} exists but is not continuous on W a solution of (2.1) cannot depend continuously on the righthand side w (<u>instability</u>). We are primarily interested in the case if well-posedness is violated by instability and therefore we shall discuss mainly the difficulties which arise in connection with instability.

If the problem (2.1) is ill-posed one could try to restore well-posedness by changing the spaces U,W and their topologies. But this approach is inadequate in the most cases since the data space and its topology is determined by practical needs: The space W and its topology must be suitable for describing any result of a measurement. As already mentioned in section 1.3 the main idea common to the most available methods for the treatment of problems which are not stable, is to restrict the class of admissible solutions by means of a suitable a-priori-knowledge: <u>Restoration of continuity (stability) by using a -</u>

<u>priori knowledge</u>. The relevance of compactness as a way for restricting the class of admissible solutions was emphasized by TIKHONOV. The following standard theorem on compactness explains this idea; in the linear case we come back to this point in the next section.

<u>Theorem 2.3</u>

Let U and W are subsets of normed linear spaces X and Y respectively and let $G: U \longrightarrow W$ be a continuous mapping. If G is bijective and U is compact, then $G^{-1}: W \longrightarrow U$ is continuous.

<u>Proof:</u> Let $w \in W$ and $(w_n)_{n \in \mathbb{N}}$ be a sequence in W with $w = \lim_n w_n$. We set $u_n := G^{-1}(w_n), n \in \mathbb{N}$, and $u := G^{-1}(w)$. Then we have to show $u = \lim_n u_n$. Since U is a compact subset of the normed space X there exists a convergent subsequence $(u_{n_k})_{k \in \mathbb{N}}$ of $(u_n)_{n \in \mathbb{N}}$ such that $z := \lim_k u_{n_k} \in U$. From the continuity of G and the fact that G is injective we obtain

$$\lim_k G(u_{n_k}) = G(z) = w \ , \quad z = u.$$

This shows that the whole sequence $(u_n)_{n \in \mathbb{N}}$ converges to u. $\square$

2.2 Restoration of continuity in the linear case

We consider the stability problem in the linear case and restrict our study to problems in which the input space X and the data space Y are Hilbert spaces. Many of the ideas and results can be extended easily to the case of Banach spaces. Let X and Y are Hilbert spaces and let A be a bounded linear operator from X into Y. Our aim is to solve the operator equation

$$(2.3) \qquad Ax = y \qquad (x \in X , y \in Y)$$

which corresponds to equation (2.1). The continuity of A^{-1} if it exists is characterized by the following

Theorem 2.4

Let X and Y be Hilbert spaces and suppose that $A \in B(X,Y)$ is injective. Then the following conditions are equivalent:

a) $R(A)$ is closed

b) $A^{-1} : R(A) \longrightarrow X$ is continuous.

Proof: Let $W := R(A)$.

a) $\Rightarrow$ b) W may be considered as a Hilbert space. From the assumptions it is clear that the operator $\tilde{A} : X \longrightarrow W$ defined by

$$\tilde{A}x := Ax , \quad x \in X,$$

is bounded, continuous and bijective. From the theorem of Banach (see e.g. [102,p.94]) we obtain that $\tilde{A}^{-1} : W \longrightarrow X$ is continuous. This implies the continuity of A^{-1}.

b) $\Rightarrow$ a) Since A^{-1} is continuous and $X = R(A^{-1})$ is closed $R(A) = (A^{-1})^{-1}(X)$ is closed too. $\qquad\qquad\square$

Example 2.5

Let $x \in L_2(-\pi,\pi)$ be given and consider x as a function on the circle with radius r = 1. According to Poisson's formula, if a harmonic function equals x on the unit circle, then in the unit disk it is given by

$$u(r,\varphi) := \frac{1}{2\pi} \int_{-\pi}^{\pi} \frac{1 - r^2}{1+r^2-2r\cos(\varphi-\alpha)} \, x(\alpha)\,d\alpha, \quad 0 \leq r < 1, \quad -\pi < \varphi \leq \pi.$$

The problem of harmonic continuation consists in finding x from the restriction $y := u(r,.)$, $0 < r < 1$ given, that is, in solving $Ax = y$, y given, where

$$(Ax)(\varphi) := \frac{1}{2\pi} \int_{-\pi}^{\pi} \frac{1 - r^2}{1+r^2-2r\cos(\varphi-\alpha)} \, x(\alpha)\,d\alpha, \quad -\pi < \varphi \leq \pi.$$

Since

$$\left| \frac{1-r^2}{1-2r\cos(\varphi-\alpha)+r^2} \right| \leq \frac{1+r}{1-r} \quad \text{for all} \quad \alpha \in (-\pi,\pi)$$

the mapping A can be considered as a bounded operator from

$X := L_2(-\pi,\pi)$ into $Y := L_2(-\pi,\pi)$. We know that the system

$$\cos(n\Psi) \ , \ \sin(n\Psi) \ , \ n \in \mathbb{N} \cup \{0\}$$

forms an orthogonal complete system in $L_2(-\pi,\pi)$.
Therefore, if x has the Fourier series

$$x(\varphi) = \sum_{j=0}^{\infty} (\alpha_j \cos(j\varphi) + \beta_j \sin(j\varphi))$$

Ax has the Fourier series

$$(Ax)(\varphi) = \sum_{j=0}^{\infty} r^j \{\alpha_j \cos(j\varphi) + \beta_j \sin(j\varphi)\}$$

The ill-posedness of the equation (*) now becomes clear from
the following identities:

$$\|x_n\|^2 = \pi \ , \ \|Ax_n\|^2 = \pi r^{2n} \ , \ \lim_n \|Ax_n\| = 0$$

$$\text{for } x_n(\varphi) := \cos(n\varphi) \hspace{4cm} *$$

The theorem above shows that the problem to solve the equation (2.3) is ill-posed if $R(A)$ is not closed. The following theorem shows that this situation is typical if the operator A is compact. Recall that a bounded operator is called compact if the image of every bounded set is relatively compact (see also Chapter 4).

Theorem 2.6

Let X and Y be Hilbert spaces and suppose that $A \in B(X,Y)$ is injective and compact. Then if A is not a finite rank operator, that is if dim $R(A) = \infty$, $R(A)$ is not closed.

Proof: Let $W := R(A)$. Assume that W is closed. Theorem 2.4 implies that $A^{-1} : W \longrightarrow X$ is continuous and therefore the identity operator

$$I = AA^{-1} : W \longrightarrow W$$

is compact too. From the well-known fact that the unit ball in a normed space is compact if and only if the dimension of the space is finite we obtain dim $R(A) < \infty$.
$\square$

If the equation (2.1) is ill-posed due to the lack of stability we may try to restore the stability by restricting the class of admissible solutions. The following theorem gives a hint how to do this if the operator A is compact.

Theorem 2.7

Let A be a compact operator from X into Y which has an inverse A^{-1}: $R(A) \longrightarrow X$ and let K be a closed convex subset of X. Then the following conditions are equivalent:

a) A^{-1} : $A(K) \longrightarrow X$ is continuous.

b) $K \cap B_r$ is compact for all $r > 0$.

__Proof:__ We set $M_r := K \cap B_r$, $N_s := A(K) \cap B_s$, $r,s > 0$.

a) $\Rightarrow$ b) Let $r > 0$. $A(M_r)$ is closed as we see as follows:

Let $(x_n)_{n \in \mathbb{N}}$ be a sequence in M_r such that $(Ax_n)_{n \in \mathbb{N}}$ converges to an element $y \in Y$. Since M_r is convex, closed and bounded it follows that M_r is weakly compact. Therefore we may assume that $(x_n)_{n \in \mathbb{N}}$ converges weakly to an element $x \in M_r$. This implies that $(Ax_n)_{n \in \mathbb{N}}$ converges weakly to Ax and therefore $Ax = y$, i.e. $y \in A(M_r)$.

Since M_r is bounded and A is compact $A(M_r) = \overline{A(M_r)}$ is compact. This implies that $M_r = A^{-1}(A(M_r))$ is compact since A^{-1} is continuous.

b) $\Rightarrow$ a)

We prove first the following assertion:

(∗) $\forall s > 0 \ \exists r > 0 \ (N_s \subset A(M_r))$.

Assume the contrary: There exists some $s_o > 0$ and a sequence $(z_n)_{n \in \mathbb{N}}$ in K such that

$$\|Az_n\| \leqq s_o \ , \quad \|z_n\| > n \ , \quad n \in \mathbb{N}.$$

Let $t := 1 + \|z_1\|$ and $u_n := \|z_n\|^{-1} z_n + (1 - \|z_n\|^{-1}) z_1 \in K$, $n \in \mathbb{N}$. Then we have $u_n \in M_t$ for all $n \in \mathbb{N}$. Since M_t is compact there exists a convergent subsequence $(u_{n_k})_{k \in \mathbb{N}}$ and $u \in M_t$ such that $u = \lim\limits_{k} u_{n_k}$. From

$$\| u_n - z_1 \| = \| z_n \|^{-1} \| z_n - z_1 \| \geq 1 - \| z_n \|^{-1} \| z_1 \| \quad , \quad n \in \mathbb{N},$$

follows $u \neq z_1$. This is a contradiction to

$$Au = \lim\limits_{k} Au_{n_k} = Az_1 .$$

Now we prove the implication b) $\Rightarrow$ a). Theorem 2.3 implies that $A^{-1} : A(M_r) \longrightarrow X$ is continuous for each $r > 0$. The assertion (*) shows that $A^{-1} : N_s \longrightarrow X$ is continuous for each $s > 0$, hence $A^{-1} : A(K) \longrightarrow X$ is continuous. $\qquad\Box$

For practical purposes mere continuous dependence of the solution on the data is not sufficient since the continuity may be arbitrary weak. A measure for the continuity of

$$(2.4) \qquad A^{-1} : A(K) \longrightarrow X$$

is given by the modul of continuity:

$$(2.5) \qquad \omega_K(\rho) := \sup \{ \| u - v \|_X \mid \| A(u - v) \|_Y \leq \rho, \ u,v \in K \}, \rho > 0.$$

One distinguishes two types of continuity:[*]

$$\text{Hölder continuity: } \omega_K(\rho) = O(\rho^\alpha) \ , \ \alpha \in (0,1].$$

$$\text{Logarithmic continuity: } \omega_K(\rho) = O(|\ln\rho|^{-\alpha}), \ \alpha > 0.$$

It is clear that for numerical reasons it is desirable to restore the continuity by means of a-priori-restrictions in such a way that Hölder continuity results. Unfortunately not in all ill-posed problems Hölder continuity can be achieved. We shall become acquainted with examples for each type of continuity.

---

[*] We denote by $O(\cdot)$ and $o(\cdot)$ the Landau-symbols.

2.3 Stability estimates

We consider again the linear equation (2.3). In order to estimate the modul of continuity ω_K we have to specify the restriction set K. We do this in the following way:

Let V and Z are Hilbert spaces and let $B \in B(V,Z)$. Assume that following conditions hold:

   i)   V is a dense subspace of X and the imbedding
        of V into X is continuous

   ii)   B is surjective.

We set

$$(2.6) \qquad K := K_E := \{v \in V \mid \|Bv\|_Z \leq E\} \, , \quad E > 0.$$

This method to specify a restriction set K is called <u>restoration of continuity by a-priori bounds</u> . In "physical" applications the bound E in (2.6) might represent an upper bound on energies, temperatures or velocities.

Remark 2.8

Another method to specify a restriction set K is to choose a cone C in Z and to define K by

$$K := K_+ := \{v \in V \mid Bv \in C\}.$$

This method is called <u>restoration of continuity by descriptive constraints</u>. The denotation and notation comes from concrete realizations of K_+ : K_+ may be for example a set of functions which are monotone and/or convex. In the light of Theorem 2.3 and Theorem 2.7 it is immediately clear that the handling of this method is more difficult since a cone is far from being compact. We shall discuss almost exclusively the proposal (2.6).   *

It is convenient to modify the definition of ω_K a little bit:

$$(2.7) \quad \omega(\rho,E) := \sup\{\|u\|_X \mid \|Au\|_Y \leq \rho, u \in V, \|Bu\|_Z \leq E\}, \rho > 0, E > 0.$$

Since we have the identities

$$\omega(\rho,E) = E\omega(\rho E^{-1},1) = \rho\omega(1,E\rho^{-1})$$

we see that Hölder continuity means that

$$\omega(\rho,E) = O(E^{1-\alpha}\rho^{\alpha}) \quad , \quad \alpha \in (0,1].$$

To prove a bound for $\omega(\rho,E)$ we need the following lemma:

<u>Lemma 2.9</u>

Let V,X,Y are Hilbert spaces, let $A \in B(X,Y)$ be injective and suppose that V is compact imbedded in X, that is the imbedding of V into X is compact. Then

$$\forall \delta > 0 \quad \exists c(\delta) > 0 \quad \forall v \in V(\|v\|_X \leq \delta\|v\|_V + c(\delta)\|Av\|_Y).$$

<u>Proof:</u> Assume the contrary: There exists $\delta > 0$ and sequences $(c_n)_{n\in\mathbb{N}}$ and $(v_n)_{n\in\mathbb{N}}$ in $\mathbb{R}$ and V respectively such that

$$(*) \qquad c_n > 0, \quad \|v_n\|_X > \delta\|v_n\|_V + c_n\|Av_n\|_Y \ , \ n \in \mathbb{N} \ ,$$
$$\lim_n c_n = \infty.$$

Then we have $\|v_n\|_V \neq 0$ and we may define $u_n := \|v_n\|_V^{-1}v_n$, $n \in \mathbb{N}$. From $(*)$ we obtain

$$(\#) \qquad \|u_n\|_X > \delta + c_n\|Au_n\|_Y \geq \delta > 0, \quad \|u_n\|_V = 1, \quad n \in \mathbb{N}.$$

Since the imbedding of V into X is bounded the sequence $(u_n)_{n\in\mathbb{N}}$ is bounded in X. This implies by $(\#)$

$$(**) \qquad \lim_n \|Au_n\|_Y = 0.$$

Since the ball B_1 in V is weakly compact and since the imbedding of V into X is compact we may assume (without loss of generality) that there exists $u \in V$ such that

$$u = w - \lim_n u_n \qquad \text{(weak convergence in V)}$$
$$u = \lim_n u_n \qquad \text{(in X)}$$
$$Au = \lim_n Au_n \qquad \text{(in Y)}.$$

From $(**)$ we obtain $Au = 0$ and this implies $u = \theta$ since A is one-to-one. But from ($\ddagger$) we know $\|u\|_X \geq \delta > 0$. This is a contradiction.

$\square$

It is easy to see that the function $\delta \longmapsto c(\delta)$ whoose existence is proved in Lemma 2.9 may be assumed to be non-increasing. If the inverse of A is not bounded then this function cannot be bounded in a neighborhood of $\delta = 0$. A typical behavior of $c(\delta)$ in this case is given by $c(\delta) = O(\delta^{-r})$, $r > 0$, as we shall see in Section 5.1.

Let us collect the assumptions which we need to prove a bound for the modul $\omega(E, \rho)$:

(2.8)   i) $A \in B(X,Y)$ and A injective.

ii) The imbedding of V into X is compact.

iii) The function
$$(0, \infty) \ni \delta \longmapsto c(\delta) \in (0, \infty)$$
is continuous and non-increasing.

iv) The norm $\|\cdot\| := \|\cdot\|_X + \|B\cdot\|_Z$ is equivalent to the norm $\| \ \|_V$ in V.

As a consequence of (2.8)iii) we obtain that the function
$$(0, \infty) \ni \delta \longmapsto \frac{\delta}{c(\delta)} \in (0, \infty)$$

is continuous, increasing and has an inverse function γ which satisfies

(2.9) $\lim\limits_{\mu \to 0} \gamma(\mu) = 0$, $\lim\limits_{\mu \to \infty} \gamma(\mu) = \infty$.

Theorem 2.10

Let the assumptions (2.8) are satisfied. Then there exist constants c, $\sigma > 0$ such that for all $\rho, E > 0$ with $\frac{\rho}{E} \leq \sigma$

$$\omega(\rho, E) \leq c \ E\gamma\left(\frac{\rho}{E}\right).$$

<u>Proof:</u>   Let $u \in V$ with $\|Au\|_Y \leq \rho$ and $\|Bu\|_Z \leq E$. From Lemma 2.9 we have for all $\delta > 0$

$$\|u\|_X \leq \delta\|u\|_V + c(\delta)\|Au\|_Y$$

$$\leq \delta c_1(\|u\|_X + \|Bu\|_Z) + c(\delta)\rho$$

$$\leq c_1\delta\|u\|_X + c_1\delta E + c(\delta)\rho$$

where $c_1 \geq 1$ is a constant such that

$$\|v\|_V \leq c_1(\|v\|_X + \|Bv\|_Z) \quad \text{for all } v \in V$$

(see (2.8)iv). Let $\sigma > 0$ be chosen such that

$$\mu := \gamma(\sigma)c_1 < 1.$$

Then we have for ρ, E with $\frac{\rho}{E} \leq \sigma$ and $\delta := \gamma(\frac{\rho}{E})$

$$(1-\mu)\|u\|_X \leq c_1\gamma(\tfrac{\rho}{E})E + \gamma(\tfrac{\rho}{E})E = (1+c_1)E\gamma(\tfrac{\rho}{E})$$

and

$$\|u\|_X \leq \frac{(1+c_1)}{1-\mu} \, E\gamma(\tfrac{\rho}{E}).$$

This shows $\omega(\rho, E) \leq \dfrac{1+c_1}{1-\mu} \, E\gamma(\tfrac{\rho}{E}).$

$\square$

Remarks 2.11

1) In the special case $V = Z$, $B = I$ we have $c = 1$ and $\sigma = \infty$ in Theorem 2.10.

2) Using the singular value decomposition for compact operators which we shall prove in Chapter 4 we can specify the function γ more precisely.

Bibliographic comments

Almost all material presented in this chapter is standard in the context of ill-posed problems; the interested reader is referred to BERTERO, VIANO, MOL [12], FRANKLIN [36] and MILLER [73]. The result of Theorem 2.6 is taken from BRYNIELSON [13]. Some heuristic considerations about restoration by descriptive constraints can be found in BUTLER, REEDS [15] and SABATIER [89].

Chapter 3: Regularization

In this section we study the problem to reconstruct the solution of an ill-posed equation from non-exact data.

3.1 Reconstruction from non-exact data

From now we fix throughout this chapter the following setting (the meaning of the objects becomes clear in the sequel; see also Section 2.3):

Let V,X,Y,Z are Hilbert spaces and let $x^{o} \in X$, $y^{o},y^{\varepsilon} \in Y$. Suppose that the following conditions are satisfied:

i) $V \subset X$ with a dense and compact imbedding;

(3.1)   ii) $A \in B(X,Y)$ injective, $R(A)$ dense in Y;

iii) $B \in B(V,Z)$ surjective.

(Since in the most cases it is clear what norm we use we omit in the following mostly the sign of space in the notations of norms and inner products).

We consider the equation

(3.2) $Ax = y.$

If $y^{o} \in R(A)$ then there exists a unique $x^{o} \in X$ such that $Ax^{o} = y^{o}$. In experimental situations the data y^{o} is not exact known due to measurement error and/or incompleteness. In a simple additive model for the perturbation of the data we may assume that for y^{o} a distorted data $y^{\varepsilon} \in Y$ is available satisfying

$$\| y^{o} - y^{\varepsilon} \| \leq \varepsilon$$

where $\varepsilon \geq o$ is the so-called noise level. The reconstruction problem consists in finding a reasonable approximation x^{ε} for x^{o} using the data y^{ε}. Since we may not assume $y^{\varepsilon} \in R(A)$ we cannot define x^{ε} as the solution of (3.2) for $y = y^{\varepsilon}$. But it is reasonable to reformulate the reconstruction problem in the following way:

Find $x^\varepsilon \in X$ which satisfies

$$(3.3) \qquad \| Ax^\varepsilon - y^\varepsilon \| \leq \varepsilon$$

If $x^\varepsilon \in X$ satisfies (3.3) we see that the defect $Ax^\varepsilon - y^\varepsilon$ has the same order as the error $y^\varepsilon - y^o$. However, if A^{-1} is not bounded, the set

$$M(\varepsilon) := \{x \in X \,\big|\; \| Ax - y^\varepsilon \| \leq \varepsilon\}$$

is not bounded and we may expect that an element x^ε which satisfies only (3.3) is in general no good approximation for x^o. In order to shrink the "solution set" $M(\varepsilon)$ we introduce again a-priori restrictions. Obviously, such a restriction set K should have the property that the mapping

$$(3.4) \qquad A^{-1} : A(K) \longrightarrow X \quad \text{is continuous and that}$$

$$(3.5) \qquad x^o \in K.$$

If the set K is chosen according to (3.4),(3.5) then we may formulate the reconstruction problem in the following way:

Find $x^\varepsilon \in X$ such that

$$(3.6) \qquad x^\varepsilon \in K, \ \| Ax^\varepsilon - y^\varepsilon \| \leq \varepsilon$$

Since $x^o \in K$, the "solution set"

$$M_K(\varepsilon) := \{x \in X \,|\; x \in K, \ \| Ax - y^\varepsilon \| \leq \varepsilon\}$$

is not empty. We are interested in an estimation of $\| x^\varepsilon - x^o \|_X$, where $x^\varepsilon \in M_K(\varepsilon)$ is arbitrary. The worst case is described by

$$\| x^\varepsilon - x^o \|_X \leq \sigma_K(\varepsilon) := \sup\{ \| x^1 - x^2 \|_X \,\big|\; x^1, x^2 \in M_K(\varepsilon) \}.$$

If we restrict us to the case

$$K := K_E := \{v \in V \,\big|\; \| Bv \| \leq E\}$$

(see Section 2.3) then we have

$$\sigma(\varepsilon,E) \;:=\; \sigma_{K_E}(\varepsilon) \;=\; \sup\{\|x^1 - x^2\|_X \,\big|\, x^i \in V, \|Bx^i\| \leq E,$$
$$\|Ax^i - y^\varepsilon\| \leq \varepsilon, i = 1,2\}$$

The following corollary shows that there is a simple connection of the quantity $\sigma(\varepsilon,E)$ with the modul of continuity $\omega(\varepsilon,E)$ defined already in (2.7).

Corollary 3.1

We have
$$\sigma(\varepsilon,E) \;\leq\; 2\omega(\varepsilon,E) \;=\; 2E\omega(\varepsilon E^{-1},1) \;.$$

Proof: $\sigma(\varepsilon,E) \leq \sup \{\|x^1 - x^2\| \,\big|\, \|Ax^1 - Ax^2\| \leq 2\varepsilon, x^i \in V, \|Bx^i\| \leq E,$
$$i = 1,2\}$$
$$\leq \sup \{\|x^1 - x^2\| \,\big|\, \|A(x^1 - x^2)\| \leq 2\varepsilon, \|B(x^1 - x^2)\| \leq 2E,$$
$$x^{1,2} \in V\}$$
$$= \omega(2\varepsilon,2E) \;=\; 2\omega(\varepsilon,E) \,.$$

$$\square$$

Example 3.2

Let us consider again the problem of harmonic continuation (see Ex.2.5)

$$X \;:=\; Y \;:=\; L_2(-\pi,\pi) \;,\quad 0 < r < 1 \;,$$

$$A : X \longrightarrow Y, \quad (Ax)(\varphi) \;:=\; \frac{1}{2\pi} \int_{-\pi}^{\pi} \frac{1 - r^2}{1 - 2r\,\cos(\varphi - \psi) + r^2} x(\psi)d\psi$$

We choose $V := H^1(-\pi,\pi)$, $Z := V$, $B : V \ni v \longmapsto v \in V$.

By Corollary 3.1 we have an estimate for $\sigma(\varepsilon,E)$, if we can estimate

$$\omega(\varepsilon,1) \;:=\; \sup\{\|x\|_X \,\big|\, \|Ax\|_Y \leq \varepsilon, \|x\|_V \leq 1\}, \varepsilon > 0 \,.$$

Let $x \in V$ be given with $\|Ax\| \leq \varepsilon$ and $\|x\|_V \leq 1$.

Then (see Ex.2.5)

$$x(\varphi) = \sum_{j=o}^{\infty} \alpha_j \cos(j\varphi) + \beta_j \sin(j\varphi)$$

$$\|x\|_X^2 = 2\pi\alpha_o^2 + \pi \sum_{j=1}^{\infty} (\alpha_j^2 + \beta_j^2)$$

$$\|Ax\|_X^2 = 2\pi\alpha_o^2 + \pi \sum_{j=1}^{\infty} r^{2j}(\alpha_j^2 + \beta_j^2)$$

$$\|x\|_V^2 = 2\pi\alpha_o^2 + \pi \sum_{j=1}^{\infty} (1 + j^2)(\alpha_j^2 + \beta_j^2).$$

Let ε be small and let $j = j(\varepsilon)$ be the unique solution of

$$(1 + j^2) = r^{2j}\varepsilon^{-2} .$$

We obtain

$$\|x\|_X^2 = 2\pi\alpha_o^2 + \pi \sum_{j\leq j(\varepsilon)} r^{2j}r^{-2j}(\alpha_j^2 + \beta_j^2)$$

$$+ \pi \sum_{j>j(\varepsilon)} (1 + j^{-2})(1 + j^2)^{-1}(\alpha_j^2 + \beta_j^2)$$

$$\leq r^{-2j(\varepsilon)}\{2\pi\alpha_o^2 + \sum_{j=1}^{\infty} r^{2j}(\alpha_j^2 + \beta_j^2)\}$$

$$+ (1 + j(\varepsilon)^2)^{-1}\|x\|_V^2$$

$$\leq 2(1 + j(\varepsilon)^2)^{-1}$$

Since

$$\ln(1 + j(\varepsilon)^2) + 2j(\varepsilon)\ln\frac{1}{r} = 2\ln\frac{1}{\varepsilon}$$

and

$$\ln(1 + j(\varepsilon)^2) = o(j(\varepsilon))$$

we find

$$j(\varepsilon) \quad = O(\ln\frac{1}{\varepsilon}) .$$

This shows

$$\omega(\varepsilon,1) = O((\ln\frac{1}{\varepsilon})^{-1}) \qquad\qquad *$$

The quantity

$$SNR := \frac{E}{\varepsilon}$$

which governs the stability estimate in corollary 3.1 is called the <u>signal-to-noise-ratio</u>. It indicates how well the exact solution shows up in the perturbed data.

To solve the reconstruction problem (3.6) we have to specify a method which chooses an element $x^\varepsilon \in X$ satisfying

$$x^\varepsilon \in K, \quad \|Ax^\varepsilon - y^\varepsilon\| \leq \varepsilon.$$

In the case $K = K_E$ two methods come into mind immediately:

(3.7)   <u>The method of residuals:</u>

Let $x^\varepsilon \in V$ be that element that minimizes

$$\|Bx\|$$

with respect to the constraint $\|Ax - y^\varepsilon\| \leq \varepsilon$.

(3.8)   <u>The method of quasisolutions:</u>

Let $x^\varepsilon \in V$ be that element that minimizes

$$\|Ax - y^\varepsilon\|$$

subject to the constraint $\|Bx\| \leq E$.

Theorem 3.3

Suppose that $\|Bx^o\| \leq E$. Then if $x^\varepsilon \in X$ is constructed according to (3.7) or (3.8) the following stability estimate is valid:

$$\|x^\varepsilon - x^o\| \leq \sigma(\varepsilon,E) \leq 2\omega(\varepsilon,E).$$

<u>Proof:</u> If x^ε solves (3.7) then from

(*) $\|Bx^o\| \leq E, \quad \|Ax^o - y^\varepsilon\| \leq \varepsilon$

it follows

$$\|Bx^\varepsilon\| \leq E.$$

If x^{ε} solves (3.8) then we obtain from (*)

$$\|Ax^{\varepsilon} - y^{\varepsilon}\| \leq \varepsilon$$

Thus, we have in each case

$$\|Bx^{\varepsilon}\| \leq E, \quad \|Ax^{\varepsilon} - y^{\varepsilon}\| \leq \varepsilon$$

which implies

$$\|x^{\varepsilon} - x^{o}\| \leq \sigma(\varepsilon, E)$$

and by Corollary 3.1 $\|x^{\varepsilon} - x^{o}\| \leq 2\omega(\varepsilon, E)$. $\square$

If we consider the problems (3.7) and (3.8) as optimization problems the theory of Lagrangian multipliers leads us to the following compromise between method (3.7) and method (3.8).

(3.9)   <u>The method of Tikhonov:</u>

Let x^{ε} be that element that minimizes

$$F(x) := \|Ax - y^{\varepsilon}\|^2 + \frac{\varepsilon^2}{E^2} \|Bx\|^2$$

in V.

<u>Theorem 3.4</u>

Suppose that $\|Bx^{o}\| \leq E$. Then if x^{ε} solves (3.9) the following inequalities hold:

i) $\|Bx^{\varepsilon}\| \leq \sqrt{2}\, E, \quad \|Ax^{\varepsilon} - y^{\varepsilon}\| \leq \sqrt{2}\, \varepsilon$

ii) $\|x^{\varepsilon} - x^{o}\| \leq 2\sqrt{2}\, \omega(\varepsilon, E)$.

<u>Proof:</u> Because of Corollary 3.1 we have only to show that the inequalities in i) hold.

Since

$$\|Bx^{o}\| \leq E \quad \text{and} \quad \|Ax^{o} - y^{\varepsilon}\| \leq \varepsilon$$

we obtain

$$\frac{\varepsilon^2}{E^2} \|Bx^{\varepsilon}\|^2 \leq F(x^{\varepsilon}) \leq 2\varepsilon^2, \quad \|Ax^{\varepsilon} - y^{\varepsilon}\|^2 \leq F(x^{\varepsilon}) \leq 2\varepsilon^2 \qquad \square$$

The consequence of Th. 3.4 is that we are sure that we lose at most a factor of $\sqrt{2}$ in the stability estimate if we use the method (3.9) instead of method (3.7) or method (3.8).

If we don't know the number $\lambda := \frac{\varepsilon}{E}$ as it is typical the case in practice we may modify the method (3.8) in the following way:

(3.10)   <u>Generalized method of Tikhonov:</u>

Let $x^{\alpha,\varepsilon}$ be that element that minimizes

$$F_\alpha(x) := \|Ax - y^\varepsilon\|^2 + \alpha\|Bx\|^2$$

in V; here α is a positive number.

Let $x^{\alpha,\varepsilon}$ be a solution of (3.10). We set

$$\varepsilon_\alpha := \|Ax^{\alpha,\varepsilon} - y^\varepsilon\| \quad , \quad E_\alpha := \|Bx^{\alpha,\varepsilon}\|$$

and

$$C := \{(a,b) \in \mathbb{R}^2 \mid \exists x \in V \text{ with } \|Ax - y^\varepsilon\| \le a, \|Bx\| \le b\}.$$

Then it can be shown that $\alpha \longmapsto \varepsilon_\alpha$ is increasing, $\alpha \longmapsto E_\alpha$ is decreasing, C is convex and the curve $\alpha \longmapsto (\varepsilon_\alpha, E_\alpha)$ is the boundary of C. Moreover, $x^{\alpha,\varepsilon}$ solves (3.7) with $\varepsilon = \varepsilon_\alpha$ and $x^{\alpha,\varepsilon}$ solves (3.8) with $E = E_\alpha$. These facts are shown in the following figure.

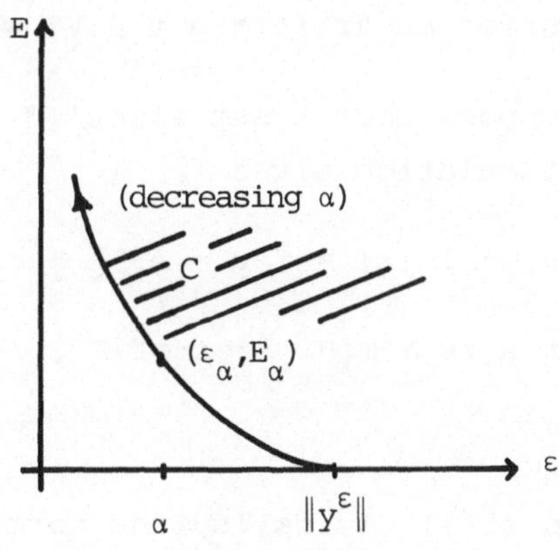

Figure 7

<u>Remark 3.5</u>

If we know that the data are exact, that is if $\varepsilon = 0$, the method (3.7) is known as a generalized BACKUS-GILBERT-method.

3.2 Preliminary results on Tikhonov's method

We consider the method (3.10) to construct an approximation $x^{\alpha,\varepsilon}$ for x^0. Let A^* denote the adjoint operator of the restriction A to the subspace V, that is

$$(Av,y)_Y = (v,A^*y)_V \text{ for all } v \in V \text{ , } y \in Y.$$

<u>Lemma 3.6</u>

A vector $x \in V$ is a minimum of F_α if and only if

$$(3.11) \qquad (A^*A + \alpha B^*B)x = A^*y^\varepsilon.$$

<u>Proof:</u> Take $v \in V$ and consider the real function
$\varphi(\lambda) := F_\alpha(x + \lambda v), \ \lambda \in \mathbb{R}.$
If x is a minimum of F_α then φ has a minimum at $\lambda = 0$, and hence

$$\begin{aligned}
0 = \varphi'(o) &= 2(Ax - y^\varepsilon, Av)_Y + 2\alpha(Bx,Bv)_Z \\
&= 2(A^*Ax - A^*y^\varepsilon,v)_V + 2\alpha(B^*Bx,v)_V.
\end{aligned}$$

Since this holds for an arbitrary $v \in V$, we have (3.11).

Conversely suppose that x satisfies (3.11). Let $u \in X$. Then a simple calculation gives

$$F_\alpha(u) - F_\alpha(x) = \|Au - Ax\|_Y^2 + \alpha\|Bu - Bx\|_Z^2$$

which shows that x is a minimizer of F_α. $\qquad\qquad\square$

The equation (3.11) is called the normal equation (associated with the functional F_α). From this equation we conclude that a sufficient condition to garantee both existence and uniqueness for the minimization problem

$$\min_{x \in V} F_\alpha(x)$$

is that the linear operator $A^*A + \alpha B^*B$ is invertible. We now introduce a condition which garantees this:

The norm

$$(3.12) \qquad \|v\|_* := (\|Av\|^2 + \|Bv\|^2)^{1/2} \quad , \quad v \in V \ ,$$

is equivalent to the given norm in V.

Theorem 3.7

Suppose that the condition (3.12) is satisfied. Then for each $y \in Y$ and each $\alpha > 0$ there exists a unique $x \in V$ satisfying

$$\|Ax - y\|^2 + \alpha\|Bx\|^2 = \min_{v \in V} \|Av - y\|^2 + \alpha\|Bv\|^2 .$$

Moreover, the element x is characterized as the unique element in V satisfying

$$(3.13) \qquad (A^*A + \alpha B^*B)x = A^*y .$$

Proof: The characterization result has been proved in Lemma 3.6, the uniqueness of x follows immediately from the normal equation (3.11) by using condition (3.12):

$$\|(A^*A + \alpha B^*B)x\| \ \|x\| \geq ((A^*A + \alpha B^*B)x,x)$$

$$= \|Ax\|^2 + \alpha\|Bx\|^2$$

$$\geq \min(1,\alpha)(\|Ax\|^2 + \|Bx\|^2)$$

$$\geq c\min(1,\alpha)\|x\|_V^2$$

Let us now prove the existence of x. Let $(x^n)_{n \in \mathbb{N}}$ be a sequence in V satisfying

$$\lim_n F_\alpha(x^n) = \inf_{v \in V} F_\alpha(v) =: \rho$$

where $F_\alpha(v) := \|Av - y\|^2 + \alpha\|Bv\|^2$, $v \in V$. Then the sequences $(\|Av_n\|)_{n\in\mathbb{N}}$ and $(\|Bv_n\|)_{n\in\mathbb{N}}$ are bounded. By (3.12) the sequence $(v_n)_{n\in\mathbb{N}}$ is bounded in V. Therefore, without loss of generality we may assume that there exists $x \in V$ such that

$$x = w - \lim_n v_n \ , \quad Ax = w - \lim_n Av_n \ ,$$

$$Bx = w - \lim_n Bv_n .$$

Since each norm is weakly lower semicontinuous we have

$$\|Ax - y\|^2 + \alpha\|Bx\|^2 \leqq \rho$$

$\square$

Obviously, condition (3.12) is satisfied if

$$V = X = Z \ , \quad B = I \ .$$

The following lemma gives a sufficient condition in the general case.

<u>Lemma 3.8</u>

Sufficient for (3.12) is the following condition:

$$(3.14) \qquad \exists \beta > 0 \ \forall v \in N(B) \ (\|Av\| \geqq \beta\|v\|_V) \ .$$

<u>Proof:</u> Obviously, there exists a constant c such that

$$\|v\|_* \leqq c\|v\|_V \quad \text{for all} \quad v \in V.$$

To prove a converse inequality assume, by way of contradiction, that for all $n \in \mathbb{N}$ there exists $v_n \in V$ such that

$$\|v_n\|_*^2 < \frac{1}{n^2} \|v_n\|_V^2 \ , \quad n \in \mathbb{N} \ ,$$

that is

$$\|Az_n\|^2 + \|Bz_n\|^2 < \frac{1}{n^2} \ , \quad n \in \mathbb{N} \ ,$$

where $z_n := \|v_n\|_V^{-1} v_n$, $n \in \mathbb{N}$. This implies

$$\lim_n Az_n = \theta, \quad \lim_n Bz_n = \theta.$$

Let $z_n = u_n + w_n$, $u_n \in N(B)$, $w_n \in N(B)^{\perp}$, $n \in \mathbb{N}$. Then

$$\lim_n Bw_n = \theta$$

and since $\tilde{B} := B\big|_{N(B)^{\perp}}$ is bijective we have

$$(*) \qquad \lim_n \|w_n\|_V = 0$$

and therefore also

$$\lim_n \|Aw_n\| = 0, \quad \lim_n \|Au_n\| = 0.$$

By (3.14) we have $\lim_n \|u_n\|_V = 0$ which implies together with $(*)$

$$\lim_n \|z_n\|_V = 0.$$

But this in contradiction to $\|z_n\|_V = 1$, $n \in \mathbb{N}$.

$\square$

Remark 3.9

If $\dim N(B) < \infty$ then condition (3.13) is satisfied. $\qquad *$

3.3   Regularizing schemes

The reconstruction problem already considered in Section 3.1 may be interpreted as the problem of finding a good approximation $R \in B(Y,X)$ for the inverse operator $A^{-1} : R(A) \longrightarrow X$. If $R(A)$ is not closed we cannot require $R\big|_{R(A)} = A^{-1}$ since this would imply that A^{-1} is continuous.

Definition 3.10

A <u>regularizing scheme</u> is a family of operators $(R_\alpha)_{\alpha > 0}$ in $B(Y,X)$ such that

$$(3.15) \qquad \lim_{\alpha \to 0} \|R_\alpha Ax - x\|_X = 0$$

for all $x \in X$.

Clearly, if A is bijective, then the family $R_\alpha := A^{-1}, \alpha > 0$, is a regularizing scheme. In the next chapters we shall consider various regularizing schemes in the case that A^{-1} is not bounded.

Theorem 3.11

Let $(R_\alpha)_{\alpha > 0}$ be a regularizing scheme. Then if $R(A)$ is not closed we have

    i) $(\|R_\alpha\|)_{\alpha > 0}$ is not bounded;

    ii) The convergence in (3.15) is not uniform on bounded subsets of X.

Proof: We know $A^{-1} : R(A) \longrightarrow X$ is not bounded.

i) By the principle of uniform boundedness $(\|R_\alpha\|)_{\alpha > 0}$ cannot be bounded since otherwise A^{-1} would be bounded.

ii) Assume the contrary. Then there exist constants $c(\alpha), \alpha > 0$, such that

$$\|R_\alpha Ax - x\| \leq c(\alpha)\|x\| \quad , \quad x \in X, \alpha > 0.$$

From this we obtain

$$\|A^{-1}y\| \leq \|A^{-1}y - R_\alpha y\| + \|R_\alpha y\|$$
$$= \|A^{-1}y - R_\alpha A(A^{-1}y)\| + \|R_\alpha y\|$$
$$\leq c(\alpha)\|A^{-1}y\| + \|R_\alpha\| \, \|y\| \, , \, y \in R(A), \alpha > 0,$$

which is in contradiction to the property that A^{-1} is not bounded.

$$\square$$

Let $(R_\alpha)_{\alpha > 0}$ be a regularizing scheme and let $x^o \in X$, $y^o, y^\varepsilon \in Y$ with

$$Ax^o = y^o \quad , \quad \|y^o - y^\varepsilon\| \leq \varepsilon.$$

Then we would like to reconstruct x^o from y^ε by defining the family

$$x^{\alpha,\varepsilon} := R_\alpha y^\varepsilon \quad , \quad \alpha > 0.$$

For the reconstruction error $\|x^{\alpha,\varepsilon} - x^\circ\|_X$ we have

$$\|x^{\alpha,\varepsilon} - x^\circ\| \leq \|R_\alpha y^\varepsilon - R_\alpha y^\circ\| + \|R_\alpha y^\circ - x^\circ\|$$

(3.16)
$$\leq \|R_\alpha\| \, \|y^\varepsilon - y^\circ\| + \|R_\alpha A x^\circ - x^\circ\|$$

$$\leq \varepsilon \|R_\alpha\| + \|R_\alpha A x^\circ - x^\circ\|.$$

The inequality (3.16) shows in connection with Theorem 3.11 that if the problem is ill-posed we have to develop a strategy for choosing the regularization parameter α in dependence on the "noise level" ε (and the exact solution x°). A reasonable parameter choice strategy should have the property that if the noise level tends to zero then the regularized solution $x^{\alpha,\varepsilon}$ converges to the solution provided the regularization parameter is chosen according to the parameter choice strategy (<u>convergence of regularization</u>). An example of such a strategy has been already mentioned in connection with the generalized method of Tikhonov (see (3.10)).

A parameter choice strategy can be derived from the estimate (3.16) by finding the best compromise between the terms $\|R_\alpha\|\varepsilon$ and $\|R_\alpha A x^\circ - x^\circ\|_X$. This means that we should choose that parameter α^* that minimizes the function

$$e(\alpha) := \|R_\alpha\|\varepsilon + \|R_\alpha A x^\circ - x^\circ\|.$$

Since in the case $\varepsilon > 0$ (see above)

$$\lim_{\alpha \to 0} \|R_\alpha A x^\circ - x^\circ\|_X = 0, \quad \lim_{\alpha \to 0} \|R_\alpha\|\varepsilon = \infty$$

we see that α plays the role of a trade-off-parameter between accuracy ($\|R_\alpha A x^\circ - x^\circ\|_X = 0!$) and stability ($\|R_\alpha\|\varepsilon$ small!). The qualitative behaviour of the bound $e(\alpha)$ for the total error is shown in the following Figure 8.

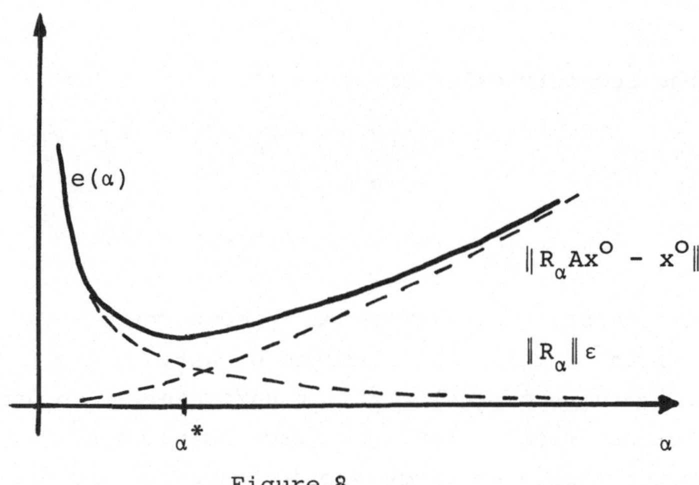

Figure 8

In practice it is difficult to adjust the parameter α in a reasonable way since x^O is not known. It is the aim of theoretical considerations to compute the asymptotics of $(\|R_\alpha\|)_{\alpha>O}$ and $(\|R_\alpha Ax^O - x^O\|)_{\alpha>O}$. To have a high order of convergence of $(\|R_\alpha Ax^O - x^O\|)_{\alpha>O}$ we have to find out reasonable assumptions on x^O.

The considerations above show that a regularizing scheme can be used to solve the equation (3.2) using the non-exact data y^ε if we can give sufficiently positive answers to the following questions:

α)   How to choose the parameter $\alpha > O$?

β)   How to compute $x^{\alpha,\varepsilon} := R_\alpha y^\varepsilon$ in a stable way ?

γ)   What is the asymptotics of $\|x^{\alpha,\varepsilon} - x^O\|$?

Thus, the description of a regularization scheme $(R_\alpha)_{\alpha>O}$ should always be completed by a sufficient analysis of the questions α,β,γ). The main difficulty is to develop reasonable strategies to choose α. We shall discuss several methods to define such strategies for various concrete regularizing schemes.

3.4 A tutorial example: The reconstruction of a derivative

In this short section we introduce a problem which may serve
as a tutorial example: reconstruction of a derivative from
measured data (see example E3) in Chapter 1 for a motivation
for this problem).

As a first step we formulate the problem in the setting of
Section 3.1:

$$X := Y := L_2(0,1) \quad \text{with the usual inner product;}$$

$$A : \quad X \longrightarrow Y \ , \quad (Ax)(t) := \int_o^t x(s)\,ds = \int_o^1 \kappa(t,s)x(s)\,ds$$

with

$$\kappa(t,s) := \begin{cases} 1 & , \ t \geq s \\ 0 & , \ t < s \end{cases} .$$

Clearly, the solution of an equation $Ax = y$ is equivalent with
the determination of $x = y'$ from y. We have:

(3.17) A is bounded with $\|A\| \leq 1$.

(3.18) A is compact.

   This is a simple consequence of the well-known
   lemma of Arzela - Ascoli.

(3.19) $R(A) = \{y \in H^1(0,1) \mid y(o) = 0\}$ is dense in Y .

(3.20) $A^* : Y \longrightarrow X \ , \ (A^*z)(s) = \int_s^1 z(t)\,dt, \ s \in (0,1).$

From the fact (3.18) it follows that the problem to solve
$Ax = y$ is ill-posed. In order to stabilize the problem we choose
(see (3.1))

$$V := H^1(0,1) \quad \text{with the usual norm} ,$$

$$Z := V \quad \text{and} \quad Bv := v, \ v \in V.$$

From the fact[*]

(3.21) $\exists c_1, c_2 \geq 0 \ \forall h > 0 \ \forall v \in V (\|v\|_X \leq c_1 h^{-1} \|v\|_V + c_2 h \|Av\|_Y)$

---

[*] See for example: Goldberg,S.:Unbounded linear operators,
McGraw-Hill,1966,p. 157 ff.

we obtain by choosing $h := \|Av\|_Y^{-\frac{1}{2}} \|v\|_V^{\frac{1}{2}}$ in (3.21) the follow-
ing

$$(3.22) \qquad \exists c \geq 0 \quad \forall v \in V(\|v\|_X \leq c\|Av\|_Y^{\frac{1}{2}} \|v\|_V^{\frac{1}{2}}).$$

This shows that

$$(3.23) \qquad \sigma(\varepsilon,E) = O(E^{\frac{1}{2}} \varepsilon^{\frac{1}{2}}).$$

The interpretation of (3.23) is that if we have $x,\tilde{x} \in V$ with
$\|x\|_V \leq E$, $\|\tilde{x}\|_V \leq E$ then

$$\|x - \tilde{x}\| \leq cE^{\frac{1}{2}} \|Ax - A\tilde{x}\|^{\frac{1}{2}}.$$

Let us define a regularizing scheme in the following way:
For $y \in Y$ we set

$$(R_h y)(t) := \begin{cases} h^{-1}(y(t+h) - y(t)) & , \quad 0 \leq t \leq \frac{h}{2} \\ h^{-1}(y(t+\frac{h}{2}) - y(t-\frac{h}{2})), & \frac{h}{2} \leq t \leq 1-\frac{h}{2} \\ h^{-1}(y(t) - y(t-h)) & , \quad 1-\frac{h}{2} \leq t \leq 1. \end{cases}$$

Let $y \in H^2(0,1)$. Using the formula

$$y(t+q) = y(t) + y'(t)q + \int_t^{t+q} (t+q-s)y''(s)ds$$

we obtain by simple arguments

$$\|R_h y - y'\| \leq h\|y''\|.$$

Therefore, given $x^o \in V$ and $y^\varepsilon \in Y$ such that $\|x^o\|_V \leq E$
and $\|y^\varepsilon - y^o\|_Y \leq \varepsilon$ then

$$\|x^o - R_h y^\varepsilon\| \leq \|x^o - R_h A x^o\| + \|R_h(y^o - y^\varepsilon)\|$$

$$\leq h E + 4h^{-1}\varepsilon$$

which implies

$$\|x^o - R_h y^\varepsilon\| \leq 5 E^{\frac{1}{2}} \varepsilon^{\frac{1}{2}}$$

if $h = \varepsilon^{\frac{1}{2}} E^{-\frac{1}{2}}$. This shows that the regularizing scheme R_h
reconstructs x^o from y^ε asymptotically with the same order as
(3.23) indicates if we choose the regularizing parameter in the
way mentioned above.

3.5 Optimal reconstruction of linear functionals

Let us consider the setting (3.1) again. In many cases one is not interested in the reconstruction of x^o itself but only in a smeared object $(u,x^o)_X$ where u is a given smearing element in X. For example this is usually the case in the reconstruction of objects from radiographs; the quantity (u,x^o) is then a local average of x^o over some given resolving length. This means that we want to reconstruct the linear functional

$$X \ni x \longmapsto (u,x) \in \mathbb{R}$$

in x^o from the data $y^\varepsilon \in Y$ using the information

$$x \in K := \{x \in V \mid \|Bx\| \leq E\}$$

and

$$y^\varepsilon \in R_\varepsilon(A) := \{y \in Y \mid \|y - Ax\| \leq \varepsilon \text{ for some } x \in X\}.$$

We assume throughout this section that

$$X \text{ and } Y \text{ are real Hilbert spaces}$$

and that

$$\varepsilon > 0 \text{ and } E = 1.$$

Definition 3.12

a) A mapping $R : R_\varepsilon(A) \longrightarrow X$ is called a <u>recovery scheme</u> and the error of R is defined as

$$\tau_R(\varepsilon) := \sup\{|(u,v) - Ry| \mid v \in K, \|Av - y\| \leq \varepsilon\}$$

b) Any recovery scheme R^* satisfying

$$\tau(\varepsilon) = \inf\{\tau_R(\varepsilon) \mid R \text{ recovery scheme}\} = \tau_{R^*}(\varepsilon)$$

is called <u>optimal.</u>

Our interest is to show that there exists an optimal linear recovery scheme. This means that there exists $w \in Y$ such that

$$R_w : Y \longrightarrow \mathbb{R} \quad , \quad y \longmapsto (w,y)_Y$$

is an optimal recovery scheme.

We need some notations:

$$e(\varepsilon) := \sup\{|(u,x)| \mid x \in K, \|Ax\| \le \varepsilon\} ,$$

$$d(\varepsilon) := \inf_{w \in Y} \sup\{|(u,x) - (w,Ax)| + \varepsilon\|w\| \mid x \in K\} ,$$

$$M := \{((u,x), Ax) \in \mathbb{R} \times Y \mid x \in K\} ,$$

$$M_d := \{(t,y) \in \mathbb{R} \times Y \mid d \le t \le d(\varepsilon), \|y\| \le \varepsilon\}.$$

The first observation is

<u>Lemma 3.13</u>

$$e(\varepsilon) \le \tau(\varepsilon) \le d(\varepsilon) .$$

<u>Proof</u>: Let R be any recovery scheme and let $x \in K$ such that $\|Ax\| \le \varepsilon$. Then we have

$$|(u,x) - R\theta| \le \tau_R(\varepsilon), \quad |(u,-x) - R\theta| \le \tau_R(\varepsilon)$$

and therefore

$$2|(u,x)| \le 2\tau_R(\varepsilon).$$

This shows $e(\varepsilon) \le \tau(\varepsilon)$.

For any $w \in Y$, $x \in K$ and $y \in Y$ satisfying $\|Ax - y\| \le \varepsilon$ we have

$$|(u,x) - (w,y)| \le |(u,x) - (w,Ax)| + \varepsilon|w|.$$

Since the mapping

$$R_w : Y \ni y \longmapsto (w,y) \in \mathbb{R}$$

is a recovery scheme for any $w \in Y$ we obtain

$$\tau(\varepsilon) \le \inf\{\tau_{R_w}(\varepsilon) \mid w \in Y\} \le d(\varepsilon). \qquad \square$$

The next lemma gives a necessary condition for the existence of an optimal linear recovery scheme.

<u>Lemma 3.14</u>

Let $d(\varepsilon) > 0$. Then the following conditions are equivalent:

a) $e(\varepsilon) = d(\varepsilon) = \tau(\varepsilon)$.

b) $M \cap M_d \neq \emptyset$ for all $d \in (0, d(\varepsilon))$.

Proof:

a) $\Longrightarrow$ b)   Let $d \in (0, d(\varepsilon))$. Then there exists $x \in K$ with

$\|Ax\| \leq \varepsilon$ such that $d \leq (u, x) \leq d(\varepsilon)$. This implies

$$((u, x), Ax) \in M_d \cap M.$$

b) $\Longrightarrow$ a)   Let x be an element in K with

$$((u, x), Ax) \in M_d \cap M.$$

Then $e(\varepsilon) \geq (u, x) \geq d$ and hence $e(\varepsilon) \geq d(\varepsilon)$. Lemma 3.13 gives the desired result.

$\square$

Theorem 3.15

We have
$$e(\varepsilon) = \tau(\varepsilon) = d(\varepsilon) \ .$$

Proof: If $d(\varepsilon) = 0$ we have nothing to prove due to Lemma 3.13. Let $d(\varepsilon) > 0$. We show that $M \cap M_d \neq \emptyset$ for all $d \in (0, d(\varepsilon))$ and apply then Lemma 3.14.

Assume that
$$M_d \cap M = \emptyset \quad \text{for some} \quad d \in (0, d(\varepsilon)).$$

From
$$M_d, M \text{ convex, closed}$$
$$\text{int } M_d \neq \emptyset \ , \ M_d \cap M = \emptyset$$

we obtain by a wellknown separation theorem that there exist $(a, w) \in \mathbb{R} \times Y$ and $b \in \mathbb{R}$ such that

(1) $(a, w) \neq (0, \theta)$,

(2) $at + (w, y) \geq b$ for all $t \in [d, d(\varepsilon)]$ and all $y \in B_\varepsilon$,

(3) $a(u, x) + (w, Ax) \leq b$ for all $x \in K$.

Setting $y := \theta$, $t := d$ in (2) gives $ad \geq b$ and setting $x := \theta$,

in (3) yields b ≥ O and therefore a ≥ O. Moreover, if a = O
then b = O and (2) implies w = θ which is impossible by (1).
Thus a > O and we may assume without loss of generality a = 1.
Then

$$\sup_{x \in K}\{(u,x) + (w,Ax)\} \leq b$$

and

$$-\varepsilon\|w\| \geq b - d$$

so that

$$d(\varepsilon) \leq \sup\{|(u,x) + (w,Ax)| + \varepsilon\|w\| \mid x \in K\}$$

$$\leq b - b + d < d(\varepsilon)$$

This is a contradiction.

□

Let $\phi_\varepsilon(y) := \sup\{(u,x) \mid x \in K, \|Ax - y\| \leq \varepsilon\}$, $y \in Y$.

Lemma 3.16

Let $w \in Y$ and $R : Y \ni y \longmapsto (w,y) \in \mathrm{IR}$. Then R is an optimal recovery scheme if and only if

$$(3.24) \qquad \phi_\varepsilon(y) \leq \phi_\varepsilon(\theta) + (w,y) \quad \text{for all} \quad y \in Y .$$

Proof: Let R be optimal, but

$$(w,y) + \phi_\varepsilon(\theta) < \phi_\varepsilon(y) \quad \text{for some} \quad y \in Y.$$

Then

$$\tau(\varepsilon) = \tau_R(\varepsilon) \geq \phi_\varepsilon(y) - (w,y) > \phi_\varepsilon(\theta) = e(\varepsilon)$$

which is a contradiction by Theorem 3.15. Suppose that (3.24)
holds and let $y \in R_\varepsilon(A)$. Then we have

$$\phi_\varepsilon(y) - (w,y) \leq \phi_\varepsilon(\theta) = e(\varepsilon)$$

$$\phi_\varepsilon(-y) + (w,y) \leq e(\varepsilon)$$

and hence

$$(u,x) - (w,y) \leq e(\varepsilon) \quad \text{for all} \quad x \in K ,$$

$$(u,x) + (w,y) \leq e(\varepsilon) \quad \text{for all} \quad x \in K.$$

This gives

$$\tau_R(\varepsilon) \leq e(\varepsilon)$$

and therefore $\tau_R(\varepsilon) = \tau(\varepsilon)$ by Lemma 3.13.　　　　　□

Theorem 3.17

Let $e(\varepsilon) < \infty$. Then there exists $w \in Y$ such that

$$R : Y \ni y \longmapsto (w,y)_Y \in \mathbb{R}$$

is an optimal recovery scheme.

__Proof:__　By Lemma 3.16 it is enough to show that

$$\phi_\varepsilon(\Theta) := \{w \in Y \mid \phi_\varepsilon(y) \leq \phi_\varepsilon(\Theta) + (w,y) \text{ for all } y \in Y\}$$

is not empty. Let

$$G := \{(y,r) \in Y \times \mathbb{R} \mid \phi_\varepsilon(y) \geq - r\}.$$

Then it is easy to see that G is convex and that each element $(y, - \phi_\varepsilon(y))$, $y \in Y$, is a boundary point of G.

Let $y \in B_{\varepsilon/2}$. Then for each $x \in K$ with $\|Ax\| \leq \varepsilon$ we have $x' := \frac{1}{2} x \in K$, $\|Ax' - y\| \leq \varepsilon$ and therefore

$$\phi_\varepsilon(y) \geq (u,x') = \frac{1}{2}(u,x) .$$

This shows

$$\phi_\varepsilon(y) \geq \frac{1}{2} \phi_\varepsilon(\Theta) .$$

Therefore the open set

$$G' := \{(y,r) \in Y \times \mathbb{R} \mid \|y\| < \frac{\varepsilon}{2} , \; r > - \frac{1}{2} \phi_\varepsilon(\Theta)\}$$

is contained in G. Thus we have the following properties:

$$G \text{ is convex, int } G \neq \emptyset ,$$

$$(\Theta, - \phi_\varepsilon(\Theta)) \text{ is a boundary } \text{point of G.}$$

By a wellknown separation theorem there exists $(w,\alpha) \in Y \times \mathbb{R}$ such that

$$(w,\alpha) \neq (\theta,0) ,$$

$$(w,y) + \alpha r \leq (w,\theta) - \alpha\phi_\varepsilon(\theta) \text{ for all } (y,r) \in G.$$

Since $G' \subset G$ α cannot be zero since otherwise we would have $(w,\alpha) = (\theta,0)$ and since

$$\sup\{(-w,y) + \alpha r \mid (y,r) \in G\} < \infty$$

we must have $\alpha \leq 0$. Thus we may assume without loss of generality $\alpha = -1$. From

$$(-w,y) + \phi_\varepsilon(y) \leq \phi_\varepsilon(\theta)$$

we conclude $w \in \partial\phi_\varepsilon(\theta)$. $\qquad\qquad\square$

Remark 3.18

In the proof above we have shown that the concave function $\phi_\varepsilon: Y \longrightarrow \mathbb{R}$ is subdifferentiable in $y = \theta$ and that each subdifferential leads to an optimal recovery scheme. $\qquad\qquad *$

Bibliographical comments

The results presented in 3.1 - 3.3 can be found in JOHN [60], FRANKLIN [36], LOCKER,PRENTER [66] and MILLER [73]. The problem of numerical differentiation is treated in CULLUM [24] and ANDERSSEN,BLOOMFIELD [2] ; in the context of optimal recovery this problem is considered by MICCHELLI [70] and GABUSHIN [37]. More on optimal recovery schemes can be found in MELKMAN, MICCHELLI [69], MICCHELLI [71], SCHARLACH [94] and MILANESE, TEMPO [72].

PART II

REGULARIZATION METHODS

In this part we consider several methods to solve ill-posed problems by regularization. Since the singular value decomposition of compact operators is useful for the construction of regularization methods for equations which are governed by compact operators we present first this decomposition and some of its applications.

Chapter 4: The singular value decomposition

The purpose of this section is to prove the singular value decomposition for compact operators in Hilbert spaces. This decomposition shows that compact operators have a simple spectrum and illustrates in a transparent way the ill-posedness of equations which are governed by compact operators (equations of the first kind).

4.1 Compact operators

Let X and Y are Hilbert spaces and let $T \in B(X,Y)$. As it is well-known, the mapping T is called compact if and only if the image of each bounded set in X is relatively compact in Y ; we write

$$T \in B_{\infty}(X,Y).$$

A diagonalization argument shows that the space $B_{\infty}(X,Y)$ of compact operators from X into Y is a closed subspace of $B(X,Y)$. Clearly, if $\dim R(T)$ is finite then T is a compact operator; call then T a finite rank operator. The following lemma is proved by simple arguments and its prove is left as an exercise to the reader.

Lemma 4.1

Let U, V, X, Y be Hilbert spaces and let $T_1 \in B(U,X), T_2 \in B(Y,V)$. If $T \in B(X,Y)$ is compact, the operator $T_2 T T_1 \in B(U,V)$ is also compact.

As an application of Lemma 4.1 it is easy to see that the adjoint T^* of $T \in B(X,Y)$ is compact if T is compact.

Theorem 4.2

Let X and Y be Hilbert spaces and let $T \in B(X,Y)$. Then the following conditions are equivalent:

a) T is compact.

b)   There exists a sequence $(T_m)_{m \in \mathbb{N}}$ in $B_\infty(X,Y)$ such that $\lim\limits_m \|T - T_m\| = 0$.

c)   There exists a sequence $(T_m)_{m \in \mathbb{N}}$ of operators of finite rank such that $\lim\limits_m \|T - T_m\| = 0$.

Proof:

a) $\Rightarrow$ b) $T_m := T, \ m \in \mathbb{N}$.

a) $\Rightarrow$ c)   Let $\varepsilon > 0$ and $M := T(B_1)$. Since $\mathrm{cl}(M)$ is compact by assumption, we can find $y_1, \ldots, y_n \in M$ such that

$$\bigcup_{i=1}^{m} B_\varepsilon(y_i) \supset M.$$

Let P_ε be the orthogonal projection of Y into $U := \mathrm{span}\{y_i \mid 1 \le i \le m \}$; set $T_\varepsilon := P_\varepsilon T$. The operator T_ε is of finite rank and from the projection property of P_ε it follows

$$\|Tx - T_\varepsilon x\| = \|Tx - P_\varepsilon Tx\| \le \inf \{\|Tx - y_i\| \mid 1 \le i \le m\} \le \varepsilon$$

for all $x \in X$. This shows $\|T - T_\varepsilon\| \le \varepsilon$. Now it is clear how to construct the sequence $(T_m)_{m \in \mathbb{N}}$.

The implications b) $\Rightarrow$ a) and c) $\Rightarrow$ a) follow from the fact that $B_\infty(X,Y)$ is a closed subspace of $B(X,Y)$.

$\square$

Let X and Y are Hilbert spaces and let $T \in B_\infty(X,Y)$. An equation

$$x + \lambda Tx = y$$

where y is a given element in Y and λ is a given nonzero number is called a (Fredholm) equation of the second kind. An equation

$$Tx = y$$

where y is a given element in Y is called a (Fredholm) equation of the first kind. We are mainly concerned with equations of the first kind, equations of the second kind are used as approximations of first kind equations.

Standard examples of compact operators are integral operators defined by

$$(4.1) \qquad (Tx)(t) := \int_{\Omega_1} \kappa(t,s)x(s)ds \quad , \quad t \in \Omega_2 ,$$

where Ω_1 and Ω_2 are measurable bounded subsets of $\mathbb{R}^{m_1}$ and $\mathbb{R}^{m_2}$ respectively and where κ is a kernel function on $\Omega_2 \times \Omega_1$.

<u>Theorem 4.3</u>

Let T be given by (4.1) and suppose that $\kappa \in L_2(\Omega_2 \times \Omega_1)$. Then T is compact.

<u>Proof:</u> As it is known, in the Hilbert spaces $L_2(\Omega_2)$ and $L_2(\Omega_1)$ there exist complete orthonormal systems $(u_j)_{j \in \mathbb{N}}$ and $(v_j)_{j \in \mathbb{N}}$ respectively. Then the system

$$w_{ij} := u_i v_j \quad , \quad i,j \in \mathbb{N} ,$$

forms a complete orthonormal system in $L_2(\Omega_2 \times \Omega_1)$. Therefore, the element κ in $L_2(\Omega_2 \times \Omega_1)$ may be represented by the series

$$\sum_{i,j \in \mathbb{N}} (\kappa, w_{ij}) w_{ij} .$$

Now define the finite rank operator $T_N : L_2(\Omega_1) \longrightarrow L_2(\Omega_2)$ by

$$(T_N x)(t) := \int_\Omega (\sum_{i,j \leq N} (\kappa, w_{ij}) w_{ij}(t,s) x(s) ds, t \in \Omega_2.$$

A simple calculation shows

$$\|Tx - T_N x\|^2 \leq (\sum_{i \geq N, j \geq N} |(\kappa, w_{ij})|^2 \|w_{ij}\|^2) \|x\|^2.$$

Since

$$\sum_{i,j \in \mathbb{N}} |(\kappa, w_{ij})|^2 = \|\kappa\|^2$$

by Parseval's identity we have

$$\lim_N \|T - T_N\| = 0 .$$

Therefore T is compact by Theorem 4.2.

$$\square$$

Examples 4.4

1) Let us consider again the problem of harmonic continuation (see Ex.2.5 and 3.2):

$$X := Y := L_2(-\pi, \pi), \quad 0 < r < 1 ,$$

$$A : X \longrightarrow Y, (Ax)(\varphi) := \frac{1}{2\pi} \int_{-\pi}^{\pi} \frac{1-r^2}{1-2r \cos(\varphi-\Psi)+r^2} x(\Psi) d\Psi.$$

A is an integral operator with kernel

$$\kappa(\varphi, \Psi) := \frac{1}{2\pi} \cdot \frac{1-r^2}{1-2r \cos(\varphi-\Psi)+r^2} , \quad \varphi, \Psi \in (-\pi, \pi).$$

Since

$$|\kappa(\varphi, \Psi)|^2 \leq \frac{1}{(2\pi)^2} \cdot \frac{(1+r)^2}{(1-r)^2} , \quad \varphi, \Psi \in (-\pi, \pi),$$

we have $\kappa \in L_2((-\pi, \pi) \times (-\pi, \pi))$. By Theorem 4.3 A is compact.

2) The problem of restoring signals that have been degraded by a perfect lowpass filter is described by a linear equation:

$$(*) \qquad \int_{-a/2}^{a/2} \frac{\sin(\Omega(t'-s'))}{\pi(t'-s')} x(s') ds' = y(t'), \quad |t'| \leq \frac{a}{2} ;$$

here Ω is the highest frequency that is allowed to pass the filter, y is the known signal and x is unknown signal. If we introduce the new variables

$$t := \frac{2t'}{a} \ , \ s := \frac{2s'}{a} \ , \ c := \frac{a\Omega}{2}$$

the equation (*) becomes an equation of the form

$$Ax = y$$

where $(Ax)(t) := \int_{-1}^{1} \frac{\sin(c(t-s))}{\pi(t-s)} x(s)ds \ , \quad |t| \le 1.$

The quantity $\int_{-1}^{1} |x(s)|^2 ds$ is usually called the energy of the (unknown) signal. Therefore it is reasonable to consider the mapping A as an operator from $X := L_2(-1,1)$ into $Y := L_2(-1,1)$. Then from the identity

$$\int_{0}^{\infty} \frac{\sin^2(bs)}{s^2} \ ds = \frac{\pi}{2} |b|$$

it follows immediately that the kernel function

$$\kappa(t,s) := \frac{\sin(c(t-s))}{\pi(t-s)} \ , \quad |t|, \ |s| \le 1$$

is in $L_2((-1,1) \times (-1,1))$. Therefore A is bounded and compact.

$$*$$

4.2 The spectrum of compact selfadjoint operators

Let X be a Hilbert space over the scalar field $\mathbb{C}$ and denote the identity map on X by I. We assume throughout $X \ne \{0\}$.

Definition 4.5

Let $T \in B(X)$. A number $\lambda \in \mathbb{C}$ is called an __eigenvalue of T__ if there exists a nonzero vector $x \in X$ (called an eigenvector associated with λ) such that $Tx = \lambda x$; the dimension of $N(T-\lambda I)$ is called its __multiplicity__.

Lemma 4.6

If $T \in B_\infty(X)$ and $\lambda \in \mathbb{C}$, $\lambda \ne 0$, then we have

i) $R(T - \lambda I)$ is closed

ii) $R(T - \lambda I) = X$ if λ is no eigenvalue of T.

<u>Proof:</u>   Let $N := N(T - \lambda I)$, $M := R(T - \lambda I)$.

Let $y \in cl(M)$ and suppose that $y = \lim_n (T - \lambda I)(x_n)$ where $(x_n)_{n \in \mathbb{N}}$ is a sequence in X. If $y_n := (T - \lambda I)(x_n)$, and $x_n = u_n + w_n$ with $u_n \in N$, $w_n \in N^{\perp}$, $n \in \mathbb{N}$, we obtain
$$y_n = (T - \lambda I)(x_n) = (T - \lambda I)(w_n), \quad n \in \mathbb{N}.$$

(1) $(w_n)_{n \in \mathbb{N}}$ is bounded.

Assume the contrary, i.e. $\lim_n \|w_n\| = \infty$.
Let $v_n := w_n \|w_n\|^{-1}$, $n \in \mathbb{N}$. Then the sequence $(v_n)_{n \in \mathbb{N}}$ is bounded, $\lim_n (T - \lambda I)(v_n) = \Theta$, and since T is compact we may assume without loss of generality that $(Tv_n)_{n \in \mathbb{N}}$ converges. This implies the convergence of $(\lambda^{-1}(\|w_n\|^{-1} y_n + Tv_n))_{n \in \mathbb{N}}$. If we set $v := \lim_n \lambda^{-1}(\|w_n\|^{-1} y_n + Tv_n)$ we obtain from $v_n \in N^{\perp}$, $\|v_n\| = 1$, $n \in \mathbb{N}$, that $v \in N$ and $\|v\| = 1$. On the other hand, we have $(T - \lambda I)(v) = \lim_n (T - \lambda I)(v_n) = \Theta$ and therefore $v \in N$. This is a contradiction. From (1) we know that the sequence $(w_n)_{n \in \mathbb{N}}$ is bounded. Since T is compact we may assume without loss of generality that $(Tw_n)_{n \in \mathbb{N}}$ converges. This implies that $w := \lim_n (T - \lambda I)(w_n)$ exists and that $y = \lim_n (T - \lambda I)(w_n) = (T - \lambda I)(w) \in M$.

Suppose now that λ is no eigenvalue of T. Let
$$X_0 := X, \quad X_{n+1} := R((T - \lambda I)^n), \quad n \in \mathbb{N}.$$

We know from the result in i) that X_1 is closed. The theorem of Banach (see e.g. [102, p.94]) implies that $(T - \lambda I)^{-1} : X_1 \longrightarrow X_0$ is bounded. By induction we obtain from
$$X_n = (T - \lambda I)^{-1}(X_{n+1}), \quad X_{n+1} \subset X_n, \quad n \in \mathbb{N},$$
the result
$$X_n \text{ is closed for all } n \in \mathbb{N}.$$

Assume $X \neq X_1 = R(T - \lambda I)$.
By induction we obtain $X_{n+1} \subsetneqq X_n$, $n \in \mathbb{N}$. If we choose for each $n \in \mathbb{N}$ an element $x_n \in X_n \cap X_{n+1}^{\perp}$ with $\|x_n\| = 1$ we obtain from the compactness of T and the fact that the sequence $(x_n)_{n \in \mathbb{N}}$ is an orthonormal sequence the existence of $\lim_n Tx_n$. This is in contradiction to

$$\|Tx_n\|^2 = \|(T - \lambda I)(x_n) + \lambda x_n\|^2$$

$$= \|(T - \lambda I)(x_n)\|^2 + |\lambda|^2 \|x_n\|^2 \geq |\lambda|^2 > 0.$$

$\square$

Definition 4.7

Let $T \in B(X)$. The set

$$\rho(T) := \{\lambda \in \mathbb{C} \mid T - \lambda I \text{ is injective}, (T - \lambda I)^{-1} \in B(X)\}$$

is called the <u>resolvent set of T</u> and the set

$$\sigma(T) := \mathbb{C} \setminus \rho(T)$$

is called the <u>spectrum of T</u>.

Theorem 4.8

Suppose $T : X \longrightarrow X$ is compact. Then we have:

i) $\sigma(T) = \{0\} \cup \{\lambda \in \mathbb{C} \mid \lambda \text{ eigenvalue of } T\}$ if $\dim X = \infty$.

ii) $\sigma(T) = \{\lambda \in \mathbb{C} \mid \lambda \text{ eigenvalue of } T\}$ if $\dim X < \infty$.

iii) Every nonzero eigenvalue has finite multiplicity.

iv) T has at most countable many eigenvalues with no accumulation point other than zero.

<u>Proof:</u> i) Let $(e_n)_{n \in \mathbb{N}}$ be an orthonormal sequence in X. Since T is compact we have $\lim_n Te_n = \Theta$. This implies $0 \in \sigma(T)$. Suppose $\lambda \in \mathbb{C}$, $\lambda \neq 0$, is no eigenvalue of T. From Lemma 4.6 we obtain $T - \lambda I : X \longrightarrow X$ is bijective and bounded. By Banach's theorem we obtain $(T - \lambda I)^{-1} : X \longrightarrow X$ is bounded; therefore $\lambda \in \rho(T)$.

ii) This follows from the fact that T is a finite rank operator if $\dim X < \infty$.

iii) Let $\lambda \neq 0$ be an eigenvalue of T. Assume $\dim N(T - \lambda I) = \infty$. If $(e_n)_{n \in \mathbb{N}}$ is an orthonormal subset of $N(T - \lambda I)$ we get from the compactness of T that $\lim_n Te_n = \Theta$.

This is in contradiction to

$$\|Te_n\| = \|\lambda e_n\| = |\lambda| \|e_n\| = |\lambda| > 0, \ n \in \mathbb{N}.$$

iv) Assume the contrary: There exists a sequence $(\lambda_n)_{n \in \mathbb{N}}$ of eigenvalues such that $\lim_n \lambda_n = \lambda \neq 0$. Let $(e_n)_{n \in \mathbb{N}}$ be an associated orthonormal sequence of eigenvectors. Again, we obtain that $(Te_n)_{n \in \mathbb{N}}$ converges which contradicts

$$\|Te_n - Te_m\|^2 = |\lambda_n|^2 + |\lambda_m|^2 \geq \tfrac{1}{2} |\lambda|^2 > 0$$

for all $m,n \in \mathbb{N}$, $m \neq n$, sufficiently large. $\qquad\square$

We recall that an operator $T \in B(X)$ is called selfadjoint if $T = T^*$. The proof of the following lemma may be left to the reader since it is the same as for a similar result in linear algebra.

Lemma 4.9

   Suppose $T : X \longrightarrow X$ is a selfadjoint operator. Then

   i) $(Tx,x) \in \mathbb{R}$ for all $x \in X$.

   ii)   The eigenvalues of T are real.

   iii)   Eigenvectors associated with different eigenvalues are orthogonal.

From Theorem 4.8 we obtain immediately that the spectrum $\sigma(T)$ of a compact operator $T : X \longrightarrow X$ is contained in the ball $B_{\|T\|}$ of $\mathbb{C}$. For a selfadjoint operator we can prove

Theorem 4.10

   Suppose $T : X \longrightarrow X$ is a compact selfadjoint operator. Then at least one of the numbers $\|T\|$, $-\|T\|$ must be an eigenvalue.

Proof: Let $(x_n)_{n \in \mathbb{N}}$ be a sequence in X such that $\|x_n\| = 1$ for all $n \in \mathbb{N}$, $\lim_n \|Tx_n\| = \|T\|$.

Such a sequence exists by the definition of $\|T\|$. We have

$$0 \leq \|T^2 x_n - \|Tx_n\|^2 x_n\|^2 = \|T^2 x_n\|^2 - |Tx_n|^4 \leq \|T\|^2 \|Tx_n\|^2 - |Tx_n|^4$$

and since $(\|Tx_n\|)_{n\in\mathbb{N}}$ converges to $\|T\|$ we see that

$$(*) \qquad \lim_n \|T^2 x_n - \|Tx_n\|^2 x_n\| = 0.$$

T^2 as the composition of compact operators is compact and we can extract a subsequence of $(x_n)_{n\in\mathbb{N}}$, say $(x_{n_k})_{k\in\mathbb{N}}$, such that $(T^2 x_{n_k})_{k\in\mathbb{N}}$ converges to an element, say $\|T\|^2 z$. Using this in

$(*)$ we see that

$$\lim_k x_{n_k} = z, \quad T^2 z = \|T\|^2 z, \quad |z| = 1.$$

Thus, we have shown that T^2 has an eigenvalue $\|T\|^2$. From

$$(T - \|T\|I)(T + \|T\|I)z = \Theta$$

we obtain immediately that $\|T\|$ is an eigenvalue of T if $(T + \|T\|I)z = \Theta$ and $-\|T\|$ is an eigenvalue of T if $(T + \|T\|I)z \neq \Theta$. $\qquad\qquad \square$

Clearly, the operators considered in Ex.4.4 are selfadjoint.

4.3   The singular value decomposition

Let X and Y be nontrivial Hilbert spaces over the scalar field $\mathbb{C}$. Throughout this section we don't distinguish the inner products and norms in X and Y.

Lemma 4.11

Suppose $T : X \longrightarrow Y$ is a linear mapping. Then the following conditions are equivalent:

a)   T is finite rank operator.

b)   There exist a number $m\in\mathbb{N}$, an orthonormal set $f_1, \ldots, f_m$ in Y and linear independent vectors $e_1, \ldots, e_m$ in X such that

$$(4.2) \qquad Tx = \sum_{j=1}^{m} (x,e_j) f_j \quad , \quad x \in X.$$

Proof:

a) $\Rightarrow$ b) $m := \dim R(T)$. Choose orthonormal vectors $f_1,\ldots,f_m$ in $R(T)$ and set $e_j := T^* f_j$, $j = 1,\ldots,m$. Then we have for each $x \in X$

$$(*) \qquad Tx = \sum_{j=1}^{m} (f_j,Tx) f_j = \sum_{j=1}^{m} (x,e_j) f_j.$$

Since $\dim R(T) = m$ we obtain from $(*)$ that $e_1,\ldots,e_m$ are linear independent.

b) $\Rightarrow$ a) The result follows from (4.2). $\qquad\qquad\qquad\square$

Lemma 4.12

Let $(e_n)_{n\in\mathbb{N}}$ and $(f_n)_{n\in\mathbb{N}}$ are orthonormal sets in X and Y respectively and let $(\mu_n)_{n\in\mathbb{N}}$ be a nonincreasing sequence of positive numbers that converges to zero. Then the operator $T : X \longrightarrow Y$, defined by

$$(4.3) \qquad Tx := \sum_{n\in\mathbb{N}} \mu_n (x,e_n) f_n \quad , \quad x \in X \, ,$$

is a compact operator.

Proof: Let $T_k := X \longrightarrow Y$, $k \in \mathbb{N}$, be defined by

$$T_k x := \sum_{n=1}^{k} \mu_n (x,e_n) f_n \quad , \quad x \in X.$$

From

$$\|T_k x - T_l x\|^2 = \sum_{n=l+1}^{k} \mu_n^2 \, |(x,e_n)|^2$$

$$\leq \mu_{k+1}^2 \sum_{n=l+1}^{k} |(x,e_n)|^2$$

$$\leq \mu_{k+1}^2 \|x\|^2 \quad , \quad x \in X, \; l \geq k \quad ,$$

we obtain that $(T_k)_{n \in \mathbb{N}}$ is a sequence in $B(X,Y)$ which converges to T. Since each T_k is a finite rank operator we obtain the result from Theorem 4.2.

$\square$

Now we come to the spectral theorem for a compact selfadjoint operator. This theorem is a generalization of the diagonalization argument for symmetric matrices and a special case of the spectral theorem for bounded operators.

Theorem 4.13

Suppose $T \in B(X)$ is compact and selfadjoint and let $\lambda_n, n \in N$ ($N = \{1,\ldots,k\}$ or $N = \mathbb{N}$) are the distinct nonzero eigenvalues of T. Then for each $x \in X$ there exists $x_o \in N(T)$ with

$$(4.4) \qquad x = x_o + \sum_{n \in N} P_n x \,, \quad Tx = \sum_{n \in N} \lambda_n P_n x$$

where P_n is the (orthogonal) projection of X onto $N(T - \lambda_n I)$ given by

$$(4.5) \qquad P_n x = \sum_{j=1}^{m_n} (x, u_j^n) u_j^n \,, \quad x \in X \,;$$

here m_n is the multiplicity of $N(T - \lambda_n I)$ and $u_1^n, \ldots, u_{m_n}^n$ is an orthonormal set in $N(T - \lambda_n I)$, $n \in \mathbb{N}$.

Proof: We know from Th.4.8 that $m_n < \infty$ for each $n \in N$. As it is wellknown, the projection P_n exists and has a representation like (4.5). Since eigenvalues corresponding to different eigenvalues are orthogonal, the set

$$U := \bigcup_{n \in N} \{u_1^n, \ldots, u_{m_n}^n\}$$

is an orthonormal subset of X. Let $x \in M$. Then the vector

$$y := \sum_{n \in N} \sum_{j=1}^{m_n} (x, u_j^n) u_j^n$$

is welldefined since

$$\sum_{n\in N}\ \sum_{j=1}^{m_n}\ |(x,u_j^n)|^2 \le \|x\|^2$$

by Bessel's inequality and we have

$$(1) \qquad y = \sum_{n\in N} P_n x \ , \ y \in M, \quad x_o := x - y \in M^\perp.$$

Since $P_n x \in N(T - \lambda_n I)$ for all $n \in N$ we get

$$(2) \qquad Ty = \sum_{n\in N} \lambda_n P_n x.$$

We complete the proof by showing that $M^\perp \subset N(T)$ which gives
$Tx = Ty$ by (2) and the result by (2).

The properties

$$(3) \qquad T(M) \subset M \ , \quad T(M^\perp) \subset M^\perp$$

follow immediately from the fact that for each $n \in N$
$T(N(T - \lambda_n I)) \subset N(T - \lambda_n I)$. Therefore, if we define
$T_1 : M^\perp \longrightarrow M^\perp$ as the restriction of T to the subspace $M^\perp$ then
$T_1 \in B(M^\perp)$ is compact and selfadjoint. Assume that T_1 is not
the zero operator. Then T_1 has a nonzero eigenvalue λ by Th.
4.10. Clearly, λ is also an eigenvalue of T and therefore
there exists $x \in X$ with

$$(4) \qquad x \in M^\perp \ , \quad x \ne \Theta \ , \quad x \in N(T - \lambda I) \subset M.$$

This is a contradiction to $M \cap M^\perp = \{\Theta\}$. Therefore we have
$T_1 = \Theta$ and hence $M^\perp \subset N(T)$. $\qquad\qquad\qquad\qquad\square$

The next theorem is the main result of this section. For the
proof of this theorem we need the following well-known identities $(T \in B(X,Y))$:

$$(4.6) \qquad \begin{aligned} N(T) &= R(T^*)^\perp \ , & N(T)^\perp &= \overline{(R(T^*))} \ , \\ N(T^*) &= R(T)^\perp \ , & N(T^*)^\perp &= \overline{(R(T))} \ . \end{aligned}$$

Theorem 4.14
____

Suppose $A : X \longrightarrow Y$ is a compact operator. Then there exist an index set J (J = {1,...,n} or J = $\mathbb{N}$), orthonormal systems $(e_j)_{j \in J}$ and $(f_j)_{j \in J}$ in X and Y respectively and a sequence $(\sigma_j)_{j \in J}$ of real positive numbers such that the following conditions are satisfied:

   i) $(\sigma_j)_{j \in J}$ is monotone nonincreasing ,

$$\lim_j \sigma_j = 0 \text{ if } J = \mathbb{N} .$$

   ii) $Ae_j = \sigma_j f_j$, $A^* f_j = \sigma_j e_j$, $j \in J$.

   iii)   For all $x \in X$ there exists $x_o \in N(A)$ with

(4.7)
$$x = x_o + \sum_{j \in J} (x,e_j)e_j \ , \ Ax = \sum_{j \in J} \sigma_j (x,e_j) f_j \ .$$

   iv)   We have for all $y \in Y$

(4.8)
$$A^* y = \sum_{j \in J} \sigma_j (y,f_j) e_j \ .$$

<u>Proof:</u> The operator $T := A^* A : X \longrightarrow X$ is compact and selfadjoint. Therefore we may apply Th.4.13 and obtain an index set N and the sequence $(\lambda_n)_{n \in N}$ of the distinct nonzero eigenvalues of $T = A^* A$. Let $(\lambda_j)_{j \in J}$ be the sequence of the eigenvalues $(\lambda_n)_{n \in N}$ repeated according to its multiplicity and ordered such that this sequence is nonincreasing. Let $(e_j)_{j \in J}$ the associated orthonormal sequence of eigenvectors of T. Since T is nonnegative, that is $(x,Tx) \geq 0$ for all $x \in X$, each eigenvalue λ_j is positive. Therefore we may define

$$\sigma_j := \lambda_j^{1/2} \ , \ f_j := \sigma_j^{-1} Ae_j \ , \ j \in J.$$

Then $(\sigma_j)_{j \in J}$ is a nonincreasing sequence with $\lim_j \sigma_j = 0$ if $J = \mathbb{N}$ by Th.4.8. Since $(e_j)_{j \in J}$ is an orthonormal sequence $(f_j)_{j \in J}$ is an orthonormal sequence too. The identities in ii) are simple consequences of the definition of σ_j, e_j and f_j. Let $x \in X$. Since $N(T) = N(A^* A) = N(A)$ the representation

$$x = x_o + \sum_{j \in J} (x,e_j)e_j \ , \ x_o \in N(A) \ ,$$

follows from (4.4). The vector

$$Qx := \sum_{j \in J} \sigma_j (x, e_j) f_j$$

is welldefined since $\|Qx\| \leq \sigma_1 \|x\|$ and belongs to $\overline{R(A)}$. It follows that $Qx - Ax \in N(A^*) \cap \overline{R(A)} = N(A^*) \cap N(A^*)^\perp$ by (4.6) and hence $Ax = Qx$. $\qquad\qquad\square$

Remarks 4.15

1) The representation (4.7) implies that each number σ_j is an eigenvalue of A^*A. On the other hand, if a nonzero real number λ is an eigenvalue of A^*A with eigenvector e, we obtain from

$$0 = \|(A - \lambda I)e\|^2 = \sum_{j \in J} |\sigma_j^2 - \lambda|^2 |(e, e_j)|^2$$
$$+ |\lambda|^2 \|e - \sum_{j \in J} (e, e_j) e_j\|^2$$

that there exists an index $j_0 \in J$ such that

$$\lambda = \sigma_{j_0}^2 \text{ and } (e, e_j) = 0 \text{ for all } j \in J \text{ with } \sigma_j^2 \neq \lambda.$$

2) The representation (4.7) and the remark above show that we find the numbers σ_j, $j \in J$, by computing the eigenvalues of A^*A.

3) Suppose that $A : X \longrightarrow X$ is a compact and selfadjoint operator. Then the preceding remarks imply that in this case the result of Th.4.14 may be stated as follows: There exist an index set J, an orthonormal set $(e_j)_{j \in J}$ and a sequence $(\mu_j)_{j \in J}$ of real numbers with $\lim_j \mu_j = 0$ if J is infinite such that for all $x \in X$

$$x - \sum_{j \in J} (x, e_j) e_j \in N(A), \quad Ax = \sum_{j \in J} \mu_j (x, e_j) e_j \; .$$

The representation above may be used to define the concept of a continuous function of the compact selfadjoint operator A: Suppose that g is a continuous function on $[-\|A\|, \|A\|]$. Then we can define g(A) by

$$g(A)(x) := \sum_{j \in J} g(\mu_j)(x,e_j)e_j.$$

Interesting special cases are given by the functions $g(s) := s^q (q \in \mathbb{R})$. The resulting operators A^q are the fractional powers of A.   *

Definition 4.16

Let $A : X \longrightarrow Y$ be a compact operator with representation (4.7). Then the numbers σ_j, $j \in J$, are called the <u>singular values of A</u> and we say that the family $\{(\sigma_j, e_j, f_j) \mid j \in J\}$ is a <u>singular system of A</u> and that the representation (4.7) is a <u>singular value decomposition of A</u>.

Example 4.17

Let us consider the operator $A : L_2(0,1) \longrightarrow L_2(0,1)$ defined by

$$(Ax)(t) := \int_0^t x(s)ds \quad , \quad t \in [0,1].$$

As we know from Section 3.4 the operator A is compact and A^*A is given by

$$A^*Ax(r) = \int_r^1 \int_o^t x(s)dsdt \quad , \quad r \in [0,1].$$

Therefore the eigenvalue equation $A^*Ae = \lambda e$ is equivalent to

$$-e = \lambda e'' \quad , \quad e(1) = e'(o) = 0.$$

The solutions of this boundary value problem are known:

$$\lambda_j = \frac{4}{(2j-1)^2 \pi^2} \quad , \quad e_j(t) = \cos \sqrt{\lambda_j}\, t, \; j \in \mathbb{N}.$$

Therefore the singular values are $\sigma_j = \frac{2}{(2j-1)\pi}$, $j \in \mathbb{N}$.   *

Let $A \in B(X,Y)$ be compact with dim $R(A) = \infty$ and singular system (σ_j, e_j, f_j), $j \in J$. Then we know from Section 2.3 that the associated equation of the first kind is ill-posed. This follows also from the singular value decomposition

$$Ax = \sum_{j \in J} \sigma_j(x,e_j)f_j \quad , \quad x \in X ,$$

since we have

$$\|e_k\| = 1 \text{ for all } k \in J, \quad \lim_k \|Ae_k\| = \lim_k \sigma_k = 0.$$

The argument above suggests also that the faster the rate of decrease of the sequence $(\sigma_j)_{j \in J}$ is the more ill-posed the associated equation is. At this point we consider this statement as a motivation to look at the order of convergence of the sequence $(\sigma_j)_{j \in J}$. The next two sections are devoted to this question.

4.4   The min-max principle

Throughout this section let X and Y be Hilbert spaces. Recall that a selfadjoint operator $T \in B(X)$ is said nonnegative (positive) definite if

$$(Tx,x) \geq 0 \ ((Tx,x) > 0) \text{ for all } x \in X.$$

Lemma 4.18

Suppose $T : X \longrightarrow X$ is a compact selfadjoint nonnegative definite operator and let $(\lambda_j)_{j \in J}$ be the decreasing sequence of nonzero eigenvalues repeated according to its multiplicity. Then

$$\lambda_1 = \| T \|$$

and

$$\lambda_{n+1} = \inf\{\sup\{ (x,Tx) \mid x \in \operatorname{span}(x_1,\ldots,x_n)^\perp ,$$

$$\|x\| \leq 1\} \mid x_1,\ldots,x_n \in X\}.$$

Proof: $\lambda_1 = \|T\|$ by Th.4.10 and the fact that $\lambda_j \geq 0$ for all $j \in J$ due to the nonnegativity of T. Let $(e_j)_{j \in J}$ be an orthonormal system in X such that

$$Tx = \sum_{j \in J} \lambda_j (x,e_j)e_j \quad , \ x \in X$$

(see Remark 4.15,3)).

If we choose $x_j = e_j$, $1 \leq j \leq n$, then for every $x \in \operatorname{span}(x_1,\ldots,x_n)^\perp$ we have

$$(x,Tx) = \sum_{j>n} \lambda_j |(x,e_j)|^2 \leq \lambda_{n+1} \|x\|^2$$

and consequently

$$\lambda_{n+1} \geq \inf\{\sup\{(x,Tx) \mid x \in \text{span}(x_1,\ldots,x_n)^\perp,$$
$$\|x\| \leq 1\} \mid x_1,\ldots,x_n \in X\}.$$

If $x_1,\ldots,x_n \in X$ are arbitrary then there exists an
$x \in \text{span}(e_1,\ldots,e_{n+1})$ such that $\|x\| = 1$, $x \in \text{span}(x_1,\ldots,x_n)^\perp$.
We have then

$$(x,Tx) = \sum_{j=1}^{n+1} \lambda_j |(x,e_j)|^2 \geq \lambda_{n+1} \sum_{j=1}^{n+1} |(x,e_j)|^2 = \lambda_{n+1}$$

and consequently

$$\lambda_{n+1} \leq \inf\{\sup\{(x,Tx) \mid x \in \text{span}(x_1,\ldots,x_n)^\perp,$$
$$\|x\| \leq 1\} \mid x_1,\ldots,x_n \in X\}. \qquad \square$$

In the following let $(\sigma_j(T))_{j\in J}$ denote the (possible finite) nonincreasing sequence of the singular values of a compact operator $T \in B(X,Y)$ with the convention that $\sigma_j(T)$ is zero if $j \notin J$.

Theorem 4.19

Let A and Q are compact operators from X into Y. Then

   i) $\sigma_1(A) = \|A\|$.

   ii) $\sigma_{n+1}(A) = \inf\{\sup \|Ax\| \mid x \in \text{span}(x_1,\ldots,x_n)^\perp,$
$$\|x\| \leq 1\} \mid x_1,\ldots,x_n \in X\} .$$

   iii) $\sigma_{j+n+1}(A + Q) \leq \sigma_{j+1}(A) + \sigma_{n+1}(Q)$, $j = 0,1,\ldots$.

Proof: i) and ii) follow from Lemma 4.18 since each $\sigma_j(A)^2$ is an eigenvalue of $T := A^*A$. By iii)

$$\sigma_{j+n+1} \leq \inf\{\sup\{\|Ax\| + \|Qx\| \mid x \in \text{span}(x_1,\ldots,x_{n+j})^\perp,$$
$$\|x\| \leq 1\} \mid x_1,\ldots,x_{n+j} \in X\}$$

$$\leq \inf\{\sup\{\|Ax\| \mid x \in \operatorname{span}(x_1,\ldots,x_n)^\perp, \|x\| \leq 1\} \mid x_1,\ldots,x_{n+j} \in X\}$$

$$+ \inf\{\sup\{\|Qx\| \mid x \in \operatorname{span}(x_{n+1},\ldots,x_{n+j})^\perp,$$

$$\|x\| \leq 1\} \mid x_{n+1},\ldots,x_{n+j} \in X\}$$

$$= \sigma_{n+1}(A) + \sigma_{j+1}(Q). \qquad \square$$

4.5 The asymptotics of singular values

Let A be a compact operator from a Hilbert space X into a Hilbert space Y. We know that the sequence of singular values of A converges to zero if this sequence is infinite. We shall present two results concerning the rate of this convergence in the case of a selfadjoint integral operator.

Suppose that $\kappa \in L_2((-1,1)\times(-1,1))$ with $\kappa(t,s) = \kappa(s,t)$ a.e. in $(-1,1)$ and let $X := L_2(-1,1)$. Then the associated operator

$$A : X \longrightarrow X, \quad (Ax)(t) := \int_{-1}^{1} \kappa(t,s)\, x(s)\, ds \ , |t| \leq 1,$$

is a compact selfadjoint operator. If we assume that A is not a finite rank operator what we do in the sequel, the operator A has an infinite sequence $(\lambda_j)_{j \in \mathbb{N}}$ of nonzero eigenvalues which converges to zero. We assume that the sequence is ordered such that

$$|\lambda_j| \geq |\lambda_{j+1}| > 0 \ , \quad j \in \mathbb{N}.$$

If $(e_j)_{j \in \mathbb{N}}$ is the associated sequence of eigenvectors, then it can be assumed that this sequence is an orthonormal sequence. Notice: The singular values of A are $\sigma_j = |\lambda_j|$, $j \in \mathbb{N}$.

Lemma 4.20

We have:

i) $\displaystyle Ax = \sum_{j=1}^{\infty} \lambda_j (x, e_j) e_j \ , \quad x \in X.$

ii) $\displaystyle \kappa(t,s) = \sum_{j=1}^{\infty} \lambda_j e_j(t) e_j(s) \ , \quad$ a.e. in $[-1,1] \times [-1,1].$

iii) $\displaystyle \int_{-1}^{1} \int_{-1}^{1} |\kappa(t,s)|^2 \, ds\, dt = \sum_{j=1}^{\infty} \lambda_j^2 .$

<u>Proof:</u> Since $\kappa \in L_2([-1,1] \times [-1,1])$ $\quad k(\cdot) := \int_{-1}^{1} |\kappa(\cdot,s)|^2 ds$ is

integrable so that k is finite a. e. on $[-1,1]$.

Let t be a point such that $k(t) < \infty$. Then the function $\kappa(t,\cdot)$

belongs to $N(A)^{\perp}$ and due to the singular value decomposition

i) (see also Rem.4.15) we have

$$\kappa(t,s) = \sum_{j=1}^{\infty} (\kappa(t,\cdot),e_j)e_j(s) = \sum_{j=1}^{\infty} \lambda_j e_j(t)e_j(s)$$

a.e. in $[-1,1] \times [-1,1]$. Using Parseval's equality we obtain

$$\int_{-1}^{1} \int_{-1}^{1} |\kappa(t,s)|^2 ds\, dt = \sum_{j=1}^{\infty} \lambda_j^2. \qquad \Box$$

Theorem 4.21

In addition to the assumption above assume that κ is

continuously differentiable on $(-1,1)$. Then we have

$$\lambda_j = O\left(j^{-3/2}\right).$$

<u>Proof:</u> Let $n \in \mathbb{N}$ and let $[-1,1]$ be divided into n equidistant

intervals at the point $\tau_i = -1 + ih$, $i = 0,\ldots,n$, where $h := \frac{2}{n}$.

We set

$$I_i := [\tau_i, \tau_{i+1}], \quad I_{ij} := I_i \times I_j, \quad 0 \leq i,\ j \leq n-1,$$

$$\kappa_n(t,s) := \kappa(\tau_i,\tau_j) + \frac{\partial \kappa}{\partial t}(\tau_i,\tau_j)(t-\tau_i) + \frac{\partial \kappa}{\partial s}(\tau_i,\tau_j)(s-\tau_j),$$

$$\text{if } (t,s) \in I_{ij}.$$

Since κ is continuously differentiable there exists a sequence

$(\rho_n)_{n\in\mathbb{N}}$ such that

$$\max_{|t|,|s|\leq 1} |\kappa(t,s) - \kappa_n(t,s)| \leq \frac{\rho_n}{2n}, \quad n \in \mathbb{N}, \text{ and } \lim_n \rho_n = 0.$$

If we define the operator $Q_n : X \longrightarrow X$ by

$$(Q_n x)(t) := \int_{-1}^{1} \kappa_n(t,s)\, x(s) ds, \quad |t| \leq 1,$$

we see immediately that $Q_n \in B(X)$, $\dim R(Q_n) \leq n$

and

$$\int_{-1}^{1}\int_{-1}^{1}|\kappa(t,s) - \kappa_n(t,s)|^2 ds\, dt \le \frac{\rho_n^2}{n^2}\ ,\quad n \in \mathbb{N}.$$

Using L. 4.20 and the notation from Section 4.3 for the singular values $\sigma_j(T)$ of an operator T we obtain by Theorem 4.19

$$\sum_{j=n+1}^{\infty}\sigma_j(A)^2 = \sum_{j=0}^{\infty}\sigma_{j+n+1}(A)^2 \le \sum_{j=1}^{\infty}\sigma_j(A - Q_n)^2$$

$$= \int_{-1}^{1}\int_{-1}^{1}|\kappa(t,s) - \kappa_n(t,s)|^2 ds\, dt \le \frac{\rho_n^2}{n^2}\ .$$

since Q_n has only n (nonzero) singular values. This implies
$$\lim_{n} n^2\left(\sum_{j=n+1}^{\infty}\sigma_j(A)^2\right) = 0 \text{ and since}$$

$$\sum_{j=n+1}^{\infty}\sigma_j(A)^2 \ge \sum_{j=n+1}^{2n}\sigma_j(A)^2 \ge n\sigma_{2n}(A)^2$$

we obtain $\lim_{n} n^3\sigma_n(A)^2 = 0$ which is the desired result. $\square$

Next, we prove a result in the case that κ is an analytic kernel. Let $R > 1$ and let E_R denote the ellipse with foci at ± 1 and semiaxis sum R:

$$E_R := \{z \in \mathbb{C} \mid z = u+iv, \frac{u^2}{a^2} + \frac{v^2}{b^2} < 1\}\text{ with } a = \tfrac{1}{2}(R + \tfrac{1}{R}),\ b = \tfrac{1}{2}(R - \tfrac{1}{R}).$$

Theorem 4.22

Suppose that the kernel κ satisfies the following conditions:

i) κ is real, continuously differentiable and symmetric
   ($\kappa(t,s) = \kappa(s,t)$ for all $(t,s) \in [-1,1]\times[-1,1]$).

ii) $\kappa(\cdot,s)$ has an analytic extension from $[-1,1]$ to the ellipse E_R for each $s \in [-1,1]$.

iii) $|\kappa(z,s)| \le M$ for all $(z,s) \in E_R \times [-1,1]$.

Then $\lambda_j = O(R^{-j})$.

<u>Proof:</u> We show first that a function f on E_R which is analytic and bounded in E_R has an expansion in Tschebyscheff polynomials:

$$f(z) = \frac{1}{2} a_o + \sum_{n=1}^{\infty} a_n T_n(z), \quad z \in E_R, \text{ where}$$

T_n denotes the n-th Tschebyscheff polynomial,

defined recursively by

$$T_o(z) := 1, \ T_1(z) := z, \ T_n(z) := 2zT_{n-1}(z) - T_{n-2}(z), z \in \mathbb{C}.$$

Each $z \in E_R$ can be written as $z = \frac{1}{2}(w+w^{-1})$ where $R^{-1} < |w| < R$. Therefore the function $\tilde{f}$, defined by

$$\tilde{f}(w) := 2f(\frac{1}{2}(w+w^{-1})) \ , \ R^{-1} < |w| < R,$$

is analytic. This implies that this function has a Laurent expansion:

$$\tilde{f}(w) = \sum_{n=-\infty}^{\infty} a_n w^n \ , \quad R^{-1} < |w| < R,$$

where

$$a_n = \frac{1}{\pi i} \int_{|w|=r} f(\frac{1}{2}(u+u^{-1})) \ u^{-n-1} \ du \ , \ n \in \mathbb{Z}$$

for $r \in (R^{-1}, R)$. Since $a_n = a_{-n}$ for all $n \in \mathbb{Z}$ we have for all $w \in \mathbb{C}$ satisfying $R^{-1} < |w| < R$

$$2f(\frac{1}{2}(w+w^{-1})) = a_o + 2 \sum_{n=1}^{\infty} a_n T_n(\frac{1}{2}(w+w^{-1}))$$

due to the identity

$$T_n(\frac{1}{2}(w+w^{-1})) = \frac{1}{2}(w^n+w^{-n}) \ , \ w \neq 0 \ , \ n \in \mathbb{N}.$$

Hence

$$f(z) = \frac{1}{2} a_o + \sum_{n=1}^{\infty} a_n T_n(z) \ , \ z \in E_R \ , \ |a_n| \leq 2Mr^{-n} \ , n \in \mathbb{N},$$

and by a limit argument

$$|a_n| \leq 2MR^{-n} \ , n \in \mathbb{N}.$$

Now we apply the result above to the kernel κ:

$$\kappa(t,s) = \frac{1}{2} a_o(s) + \sum_{n=1}^{\infty} a_n(s) T_n(t) \quad , |t|,|s| \leq 1$$

where $|a_n(s)| \leq 2MR^{-n}, \quad n \in \mathbb{N}, \ |s| \leq 1.$

Since
$$a_n(s) = \frac{1}{\pi} \int_{-\pi}^{\pi} \kappa(\cos\varphi, s) e^{-in\varphi} d\varphi, \quad n \in \mathbb{N}, \quad |s| \leq 1,$$

we obtain that a_n is continuous for all $n \in \mathbb{N}$.

Therefore, if we define the kernel
$$q_N(t,s) := \frac{1}{2} a_0(s) + \sum_{n=1}^{N} a_n(s) T_n(t), \quad |t|, |s| \leq 1 ,$$

and the operator Q_N by
$$(Q_N x)(t) := \int_{-1}^{1} q_N(t,s) x(s) ds , \quad |t| \leq 1,$$

then
$$\dim(Q_N) \leq N + 1 ,$$

$$|\kappa(t,s) - q_N(t,s)| \leq \sum_{k=N+1}^{\infty} |a_k(s)| \leq 2M \sum_{k=N+1}^{\infty} R^{-k} \frac{2MR^{-N}}{R-1} ,$$

$$\|A - Q_N\| = O(R^{-N}).$$

From Theorem 4.19 we obtain
$$\sigma_{n+2}(A) \leq \sigma_1(A - Q_N) + \sigma_{N+2}(Q_N) = \sigma_1(A - Q_N) = \|A - Q_N\|$$

which implies the desired result.

$\square$

4.6 Picard's criterion

The main result in this section is a criterion for the solvability of linear equations which are governed by compact operators.

Theorem 4.23

Let A be a compact operator from the Hilbert space X into the Hilbert space Y and let $(\sigma_j, e_j, f_j)_{j \in J}$ be a singular system of A. Then for a given element $y \in Y$ the following conditions are equivalent:

a) $y \in R(A)$

b) $y \in \overline{R(A)}$; $\sum_{j \in J} \sigma_j^{-2} |(y, f_j)|^2 < \infty$.

<u>Proof:</u> a) $\Rightarrow$ b) We have $y \in R(A) \subset \overline{R(A)}$. Let $x \in X$ with $Ax = y$.
From Th. 4.14 we obtain

$$\sum_{j \in J} \sigma_j^{-2} |(y, f_j)|^2 = \sum_{j \in J} \sigma_j^{-2} |(x, A^* f_j)|^2 = \sum_{j \in J} |(x, e_j)|^2$$

$$\leq \|x\|^2 < \infty$$

by Bessel's inequality.

b) $\Rightarrow$ a) We have only to consider the case $J = \mathbb{N}$. If we set

$$x^n := \sum_{j=1}^{n} \sigma_j^{-1} (y, f_j) e_j , \qquad n \in \mathbb{N} ,$$

we obtain for $m, n \in \mathbb{N}$, $m \geq n$,

$$\|x^n - x^m\|^2 = \sum_{j=n+1}^{m} \sigma_j^{-2} |(y, f_j)|^2$$

which shows that $(x^n)_{n \in \mathbb{N}}$ is a Cauchy sequence.

With $x := \lim_n x^n$ we find

$$Ax = \sum_{j=1}^{\infty} (y, f_j) f_j , \qquad \|Ax\| \leq \|y\| .$$

Let us write $z := y - \sum_{j=1}^{\infty} (y, f_j) f_j$; a short calculation shows
that
$$\|z\|^2 = \|y\|^2 - \sum_{j=1}^{\infty} |(y, f_j)|^2 .$$

Also, we have $(z, f_j) = 0$ for all $j \in \mathbb{N}$, so that by the representation (4.8) $A^* z = \theta$.
Since $y \in \overline{R(A)} = N(A^*)^\perp$ this implies

$$0 = (z, y) = \|y\|^2 - \sum_{j=1}^{\infty} |(y, f_j)|^2 = \|z\|^2 .$$

Consequently

$$y = \sum_{j=1}^{\infty} (y, f_j) f_j = Ax .$$

$\square$

Remarks 4.23

1) The condition b) in Th.4.22 is called Picard's criterion. It plays a role analogous to Fredholm's alternative for equations of the second kind.

2) The proof of Th.4.22 suggests a tentative solution for the equation $Ax = y$:

$$(4.9) \qquad x = \sum_{j \in J} \sigma_j^{-1} (y, f_j) e_j.$$

This formula is very useful for theoretical considerations but it is of low importance in practical computations since the singular system is not known or is difficult to compute.

3) The singular value decomposition shows very clear the ill-posedness of linear equations which are governed by compact operators: If we perturb the data y in the equation $Ax = y$ by adding the vector $\delta y := \alpha f_k$, we obtain a perturbation $\delta x := \alpha \sigma_k^{-1} f_k$ in the solution x by formula (4.9); hence the ratio $\|\delta x\| \|\delta y\|^{-1}$ can be made arbitrary large if $J = \mathbb{N}$ due to the fact $\lim_j \sigma_j = 0$.

4) The condition b) in Th.4.22 is equivalent to the following condition:

$$b') \qquad y \in N(A^*)^\perp , \quad \sum_{j \in J} \sigma_j^{-2} | (y, f_j) |^2 < \infty.$$

Bibliographical comments

Compact operators and their spectral decompositions are discussed in SMITHIES [97], WEIDMANN [102] and GROETSCH [44]. Th. 4.21 is due to WEYL [104], a generalization is proved by LITTLE, RAEDE [65]; Th. 4.22 is a result of RAEDE [86]. See also [85].

Chapter 5: Applications of the singular value decomposition

In this chapter we demonstrate that the singular value decomposition of a compact operator plays a role in many different subjects. We restrict us to those topics which are related to problems in the next chapters.

5.1 Hilbert scales

Let H be a separable Hilbert space with inner product $(.,.)_H$ and orthonormal basis $(e_j)_{j \in \mathbb{N}}$ and let $(\alpha_j)_{j \in \mathbb{N}}$ be a sequence with

$$0 < \alpha_{j+1} \leq \alpha_j \leq 1, \ j \in \mathbb{N}, \ \lim_j \alpha_j = 0.$$

Let M be the set of elements in H representable by a finite linear combination of the elements $(e_j)_{j \in \mathbb{N}}$. Then we define on M for each $s \in \mathbb{R}$ an inner product $(.,.)_s$:

$$(x,y)_s := \sum_{j=1}^{\infty} \alpha_j^{-2s} (x,e_j)_H (y,e_j)_H.$$

(Notice that the series above is actually a finite sum). The completions H_s of M in the norm $\| \ \|_s := (.,.)_s^{1/2}$ is a Hilbert space by definition.

Lemma 5.1

We have

1) $H_o = H$, $(.,.)_o = (.,.)_H$.

2) $H_s = \{x \in H_o \mid \sum_{j=1}^{\infty} \alpha_j^{-2s} |(x,e_j)_o|^2 < \infty\}$,

$$\|x\|_s^2 = \sum_{j=1}^{\infty} \alpha_j^{-2s} |(x,e_j)_o|^2 , \ x \in H_s, \text{ for each } s \geq 0.$$

3) $H_s \subset H_t \subset H_o$ for $s \geq t \geq 0$.

4) If $s > t \geq 0$ then the imbedding of H_s into H_t is dense and compact.

5) $\|x\|_r \leq \|x\|_s^{1-\varphi} \|x\|_t^{\varphi}$ for all $x \in H_s$ if

$$s \geq r \geq t \geq 0, \quad s \neq t, \quad \text{where } \varphi = \frac{s-r}{s-t} .$$

<u>Proof:</u> The assertions 1),2) and 3) are simple consequences of the definition of H_s and $(.,.)_s$. Let us define $T_N : H_o \longrightarrow H_o$ by

$$T_N x := \sum_{j=1}^{N} (x,e_j)_o \, e_j , \quad x \in H_o.$$

Then we have

$$\|T_N x - x\|_t^2 = \sum_{j=N+1}^{\infty} \alpha_j^{-2t} |(x,e_j)_o|^2$$

$$\leq \sup_{j \geq N+1} \alpha_j^{2(s-t)} \|x\|_s^2$$

$$\leq \alpha_{N+1}^{2(s-t)} \|x\|_s^2 , \quad x \in H_s ,$$

which shows that $(T_N)_{N \in \mathbb{N}}$ converges uniformly to the imbedding of H_s into H_t. This implies that this imbedding is compact since each T_N is compact (see Th.4.2).

Let $x \in H_s$. Then

$$\|x\|_r^2 = \sum_{j=1}^{\infty} \alpha_j^{-2r} |(x,e_j)_o|^2$$

$$= \sum_{j=1}^{\infty} (\alpha_j^{-2t\varphi} |(x,e_j)_o|^{2\varphi}) (\alpha_j^{-2s(1-\varphi)} |(x,e_j)_o|^{2(1-\varphi)})$$

and by Hölder's inequality

$$\|x\|_r^2 \leq (\sum_{j=1}^{\infty} \alpha_j^{-2t} |(x,e_j)_o|^2)^{\varphi} (\sum_{j=1}^{\infty} \alpha_j^{-2s} |(x,e_j)_o|^2)^{1-\varphi}$$

$$= \|x\|_t^{2\varphi} \|x\|_s^{2(1-\varphi)} .$$

This gives the result 5). $\qquad\qquad\qquad\qquad\qquad\qquad\qquad\qquad\Box$

The spaces H_s, $s < 0$, contain generalized (ideal) elements. We clarify the structure of the spaces H_s for $s < 0$.

Let $r \in \mathbb{R}$. We define a map $\tilde{D}_r : M \longrightarrow M$ by

$$\tilde{D}_r x := \sum_{j=1}^{\infty} \alpha_j^{-r} (x,e_j)_o \, e_j \quad \text{if } x = \sum_{j=1}^{\infty} (x,e_j)_o \, e_j.$$

Notice that this definition makes sense since $\sum_{j=1}^{\infty} (x,e_j)_o \, e_j$ is a finite sum if $x \in M$. Since

$$\|\tilde{D}_r x\|_o = \|x\|_r$$

for all $x \in M$ and since M is dense in H_r the map $\tilde{D}_r$ may extended to a map $D_r : H_r \longrightarrow H_o$. The following properties of D_r are simple consequences of the definition of $\tilde{D}_r$ and the construction of D_r:

(5.1) $D_r(M) = M$.

(5.2) $\|D_r x\|_o = \|x\|_r$ for all $x \in H_r$.

(5.3) D_r is bijective.

(5.4) $D_r^{-1} : H_o \longrightarrow H_r$, $\quad D_r^{-1} x = \sum_{j=1}^{\infty} \alpha_j^{r} (x,e_j)_o \, e_j$; $\quad r \geqq 0$.

(5.5) $(D_r^{-1} u, x)_o = (u, D_{-r} x)_o$ for all $u \in H_o$, $x \in M$; $r \geqq 0$.

(5.6) $(D_{2r} x, u)_o = (D_r x, D_r u)_o$ for all $u \in H_r$, $x \in M$; $r \geqq 0$.

Theorem 5.2

The dual space H_s^* of H_s is isometric isomorph to H_{-s} for all $s \geqq 0$.

Proof: Let $s > 0$ (for $s = 0$ nothing has to be proved) and set $t := -s$.
Let $\varphi \in H_t^*$. If $x \in M$ we have

$$|<\varphi,x>_t| \leqq \|\varphi\|_{H_t^*} \|x\|_t = \|\varphi\|_{H_t^*} \|D_t x\|_o \, ,$$

$$\left|<\varphi,D_s x>_t\right| \leq \|\varphi\|_{H_t^*}\|D_s x\|_t = \|\varphi\|_{H_t^*}\|x\|_o \ ,$$

where $<.,.>_t$ is the canonic bilinear form on $H_t^* \times H_t$. This shows that the linear functional

$$M \ni x \longmapsto <\varphi,D_s x>_t \in \mathbb{R}$$

is bounded in the norm of H_o on the set $D_s(M)$. Since $D_s(M) = M$ and since M is dense in H_o there exists by the Riesz representation theorem an element $u \in H_o$ with

$$<\varphi,D_s x>_t = (u,x)_o \quad \text{for all } x \in M.$$

This implies (see (5.5))

$$(1) \qquad <\varphi,x>_t = (u,D_t x)_o = (D_s^{-1}u,x)_o = (z,x)_o \quad \text{for all } x \in M$$

where

$$z := D_s^{-1}u \in H_s.$$

On the other hand, for every $z \in H_s$ the linear form

$$M \ni x \longmapsto (z,x)_o \in \mathbb{R}$$

defines a linear functional on M, bounded in the norm of H_t:

$$\left|(z,x)_o\right| = \left|(D_s z,D_t x)_o\right| \leq \|D_s z\|_o \|D_t x\|_o$$

$$= \|z\|_s \|x\|_t.$$

This functional can be extended by continuity to a functional φ in H_t^* with $\|\varphi\|_{H_t^*} \leq \|z\|_s$. We show that we have equality:

$$\|\varphi\|_{H_t^*} = \|z\|_s.$$

Since $z \in H_s = H_{-t}$ there exist a sequence $(z_n)_{n\in\mathbb{IN}}$ of elements in M such that $\lim_n \|z - z_n\|_s = 0$.

We set $u_n := D_{2s}z_n$, $n \in \mathbb{IN}$, and have (see (5.6))

$$(u_n,z)_o = (D_{2s}z_n,z)_o = (D_s z_n,D_s z)_o$$

and therefore

$$\lim_n (u_n, z)_o = \|D_s z\|_o^2 = \|z\|_s^2 .$$

For any $\varepsilon > 0$ and sufficiently large $n \in \mathbb{N}$ we then obtain

$$\langle \varphi, u_n \rangle_t = (u_n, z)_o \geq (1 - \varepsilon) \|z\|_s^2$$

$$\geq (1 - 2\varepsilon) \|z_n\|_s \|z\|_s = (1 - 2\varepsilon) \|u_n\|_t |z|_s .$$

This implies the desired inequality

$$\|\varphi\|_{H_t^*} \geq \|z\|_s .$$

Thus we have proved that H_t^* is isometric to H_s and an isometry is given by the map $D_t \circ J_o$ where J_o is the Riesz mapping from H_o^* into H_o. Since Hilbert spaces are reflexive, H_s^* is isometric isomorph to H_{-s}. $\qquad\qquad \square$

From Lemma 5.1 and Theorem 5.2 we obtain by identifying the spaces H_s^* with H_{-s}, $s \geq 0$, the following

<u>Theorem 5.3</u>

Let $r, t, s \in \mathbb{R}$. We have :

1) $H_t \subset H_s$ if $t \geq s$.

2) The imbedding of H_t into H_s is dense and compact if $t > s$.

3) $\|x\|_r \leq \|x\|_s^{1-\varphi} \|x\|_t^\varphi$ for all $x \in H_s$ if $s \geq r \geq t$, $s \neq t$, where $\varphi = \dfrac{s - r}{s - t}$.

4) $H_s^* = H_{-s}$ for all $s \in \mathbb{R}$.

5) $\|u\|_s = \sup \{ |(u,v)_o| \, | \, v \in H_{-s}, \|v\|_{-s} \leq 1 \}$, $u \in H_s$, for all $s \in \mathbb{R}$.

We give an example how to construct specific scales of Hilbert spaces.

<u>Example 5.4</u>

Let T be a compact injective operator from a Hilbert space H_o into another Hilbert space. Let $(\sigma_j, e_j, f_j)_{j \in \mathbb{N}}$ be a singular system of T. Then with the pair $((e_j)_{j \in \mathbb{N}}, (\sigma_j)_{j \in \mathbb{N}})$ a scale of Hilbert spaces can be constructed; we denote this scale by $(H_s(T))_{s \in \mathbb{R}}$. If we apply this construction to the case considered in Ex. 4.17 a scale of spaces of Sobolev type results.

*

<u>Definition 5.5</u>

A family $(H_s)_{s \in \mathbb{R}}$ of separable Hilbert spaces (with inner products $(.,.)_s$) is called a <u>Hilbert scale</u> if and only if the following properties hold:

i) $H_s \subset H_o \subset H_t$ with dense and continuous imbeddings; $s \geq o \geq t$.

ii) $H_s^* = H_{-s}$,

$$|(u,v)_o| \leq (u,u)_s^{1/2} \, (v,v)_{-s}^{1/2} \quad \text{for all } u \in H_s, \; v \in H_{-s}; \; s \in \mathbb{R}.$$

iii) $(u,u)_r \leq (u,u)_s^{2(1-\varphi)} (u,u)_t^{2\varphi}$ for all $u \in H_s$; $s \geq r \geq t$,

$$s \neq t, \quad \varphi := (s-r)(s-t)^{-1}.$$

In Ex.5.4 there is given a scale $(H_s(T))_{s \in \mathbb{R}}$ of Hilbert spaces which is a Hilbert scale due to Theorem 5.3.

5.2 Convergence of regularizing schemes

Throughout this section, let X and Y be Hilbert spaces and let $A : X \longrightarrow Y$ be a compact injective operator with singular system $(\sigma_j, e_j, f_j)_{j \in \mathbb{N}}$. We set $a := \sigma_1$ and denote the Hilbert scale constructed according to Ex.5.4 by $(H_s(A))_{s \in \mathbb{R}}$.

Let $x^o \in X$, $y^o, y^\varepsilon \in Y$ with

$$(5.7) \qquad Ax^o = y^o , \qquad \|y^o - y^\varepsilon\| \leq \varepsilon.$$

As we know from Section 4.5 the solution x^o of $Ax = y^o$ is given by

$$x^o = \sum_{j=1}^{\infty} \sigma_j^{-1} (y^o, f_j) e_j .$$

But if we know only the measured data y^ε we have to damp the terms (y^ε, f_j) corresponding to small singular values (see Rem. 4.23). This is done in the following by considering

(5.8) $\qquad x^{\varepsilon, \alpha} := R_\alpha y^\varepsilon$

where the family $(R_\alpha)_{\alpha > o}$ is defined by

$$R_\alpha y := \sum_{j=1}^{\infty} q(\alpha, \sigma_j) \sigma_j^{-1} (y, f_j) e_j \quad , \quad y \in Y ;$$

here $q : (o, \infty) \times (o, a] \longrightarrow \mathrm{I\!R}$ is a given function.

Definition 5.6

We say that the function q is a __damping function__ if and only if q satisfies the following conditions:

   i) $\quad o \leq q(\alpha, \sigma) \leq 1 \quad$ for all α, σ .

  ii) $\quad$ There exists for each $\alpha > o$ a constant $c(\alpha)$ such that

$$q(\alpha, \sigma) \leq c(\alpha) \sigma \quad \text{for all } \sigma \in (o, a].$$

 iii) $\quad \lim_{\alpha \to o} q(\alpha, \sigma) = 1 \quad$ for each $\sigma \in (o, a]$.

If q is a damping function then from the estimates

$$\| R_\alpha A x - x \|^2 = \Big\| \sum_{j=1}^{\infty} q(\alpha, \sigma_j) \sigma_j^{-1} \sigma_j (x, e_j) e_j - \sum_{j=1}^{\infty} (x, e_j) e_j \Big\|^2$$

$$= \sum_{j=1}^{\infty} | q(\alpha, \sigma_j) - 1 |^2 \, | (x, e_j) |^2 \quad , \quad x \in X ,$$

$$\| R_\alpha y \|^2 = \sum_{j=1}^{\infty} | q(\alpha, \sigma_j) |^2 \sigma_j^{-2} | (y, f_j) |^2$$

$$\leq c(\alpha) \| y \|^2 \quad , \quad y \in Y ,$$

we obtain immediately that the family $(R_\alpha)_{\alpha>0}$ is a regularizing scheme in the sense of Def.3.10. Relevant specific examples are given in

Examples 5.7

1) $\quad q(\alpha,\sigma) := \dfrac{\sigma^2}{\sigma^2+\alpha}$.

The resulting regularizing scheme is Tikhonov's regularization method (see Section 3.2).

2) $\quad q(\alpha,\sigma) := \begin{cases} 1 & , \ \sigma^2 \geq \alpha \\ 0 & , \ \text{else} \end{cases}$.

The resulting regularizing scheme is called the truncated singular value decomposition.

3) $\quad q(\alpha,\sigma) := 1 - \exp(-\dfrac{\sigma^2}{\alpha})$.

*

Since we have at hand only the perturbed data y^ε convergence of $(R_\alpha y^\varepsilon)_{\alpha>0}$ to x^o can be achieved only if we apply a reasonable parameter choice strategy $\alpha = \alpha(\varepsilon)$. A first observation is

Theorem 5.8

If $\alpha = \alpha(\varepsilon)$ can be chosen such that

$$\lim_{\varepsilon \to 0} \alpha(\varepsilon) = 0 \ , \quad \lim_{\varepsilon \to 0} c(\alpha)\varepsilon = 0,$$

then $\lim\limits_{\varepsilon \to 0} R_{\alpha(\varepsilon)} y^\varepsilon = x^o$.

Proof: We have

$$(5.9) \quad \|R_\alpha y^\varepsilon - x^o\| \leq \|R_\alpha y^o - x^o\| + c(\alpha)\varepsilon .$$

$$\|R_\alpha y^o - x^o\|^2 = \sum_{j=1}^{\infty} |q(\alpha,\sigma_j) - 1|^2 \, |(x^o,e_j)|^2 .$$

Since $\lim\limits_{\alpha \to o} q(\alpha,\sigma) = 1$ for each $\sigma \in (o,a]$ and since $|q(\alpha,\sigma) - 1| \leq 1$ for all α,σ the first term on the right-hand side of (5.9) converges to zero. $\qquad\qquad\square$

The first expression on the right-hand side of (5.9) is the regularizing error, the second term is the error due to the noise. The convergence rate of the regularizing error depends on "smoothness" properties of the (unknown) solution x^o. Such a property may be introduced by an assumption

$$x^o \in H_k(A) \ , \quad k > 0.$$

Then we can modify (5.9) in the following way:

$$(5.10) \qquad \|R_\alpha y^\varepsilon - x^o\|^2 \leq d_k(\alpha)^2 \|x^o\|_k^2 + c(\alpha)^2 \varepsilon^2$$

$$\text{where } d_k(\alpha) := \sup_{\sigma \in (o,a]} |q(\alpha,\sigma) - 1| \sigma^k \quad ,\alpha > o.$$

Illustration 5.9

Consider the case $q(\alpha,\sigma) := \dfrac{\sigma^2}{\sigma^2 + \alpha}$. Then we have

$$c(\alpha) = \frac{1}{2} \frac{1}{\sqrt{\alpha}} \ , \quad d_1(\alpha) \leq \sqrt{\alpha}, \quad d_2(\alpha) \leq \alpha.$$

This gives:

$$(*) \qquad \begin{aligned} k = 1 : \quad & \|R_\alpha y^\varepsilon - x^o\| \leq \sqrt{\alpha} \, \|x^o\|_1 + \frac{\varepsilon}{\sqrt{\alpha}} \\[2mm] k = 2 : \quad & \|R_\alpha y^\varepsilon - x^o\| \leq \alpha \, \|x^o\|_2 + \frac{\varepsilon}{\sqrt{\alpha}} \ . \end{aligned}$$

An optimal choice of the parameter α based on (*) consists in choosing

$$\alpha(\varepsilon) = O(\varepsilon) \qquad \text{if} \quad k = 1$$

and

$$\alpha(\varepsilon) = O(\varepsilon^{2/3}) \qquad \text{if} \quad k = 2.$$

*

5.3 On the use of the conjugate gradient method

The purpose of this section is to show that the conjugate gradient method may be used to solve a linear equation of the first kind even if noise is present.

Let A be a compact, symmetric and positive definite operator on the Hilbert space X and let $x^o, y^o, y^\varepsilon \in X$ with

$$(5.11) \qquad Ax^o = y^o \ , \quad \|y^o - y^\varepsilon\| \leq \varepsilon .$$

(The results which we shall prove may be applied if A is given as T^*T where T is a compact operator from the Hilbert space X into a Hilbert space Y).

Let $(\sigma_j)_{j \in \mathbb{N}}$ be the sequence of singular values of A and let $(e_j)_{j \in \mathbb{N}}$ be an orthonormal system of singular vectors of A. Then we know (see Rem.4.15):

$$Ae_j = \sigma_j e_j \ , \quad 0 < \sigma_{j+1} \leq \sigma_j \ , \quad j \in \mathbb{N}$$

$$\lim_j \sigma_j = 0.$$

The conjugate gradient method is an iterative method. Clearly, it is expected that the asymptotics of the sequence $(\sigma_j)_{j \in \mathbb{N}}$ has influence on the rate of convergence of the iteration. We describe the method for solving the equation

$$Ax = y \ .$$

Later on we shall apply the results for different right-hand sides y and different operators A.

The method of conjugate gradients is an iteration of the following form ($(.,.)$ is the inner product in X):

$$(5.12)$$

Choose $x_o \in X$, set $r_o := s_o := y - Ax_o$, continue with

i) $\quad d_k := As_k, \ \gamma_k := (s_k, r_k)(s_k, d_k)^{-1} \ ,$

ii) $\quad x_{k+1} := x_k + \gamma_k s_k, \ r_{k+1} := r_k - \gamma_k d_k \ ,$

iii) $\quad \beta_k := (r_{k+1}, d_k)(s_k, d_k)^{-1}, \ s_{k+1} := r_{k+1} - \beta_k s_k.$

The following identities are well-known and easy to establish (Notice that we may assume without loss of generality that $r_i \neq \Theta$, $i \in \mathbb{N}$).

Corollary 5.10

For all $i,j \in \mathbb{N}$ we have

1) $(r_i,r_j) = 0$ for $i \neq j$;

2) $(s_i,d_j) = 0$ for $i \neq j$;

3) $(r_j,s_j) = 0$ for $i < j$;

4) $\|r_i\|^2 (r_i,d_j)^{-1} < \gamma_i < \|s_i\|^2 (s_i,d_i)^{-1}$ if $r_i \neq \Theta$;

5) $s_i = \|r_i\|^2 \sum_{l=o}^{i} r_l \|r_l\|^{-2}$;

6) $\|s_i\|^2 \geq \|r_i\|^2$;

7) $\mathrm{span}\{s_o,\ldots,s_i\} = \mathrm{span}\{r_o,\ldots,r_i\}$

$$= \mathrm{span}\{r_o,Ar_o,\ldots,A^i r_o\}.$$

In order to prove convergence result we need the fact that the conjugate gradient method is an optimal process:

Lemma 5.11

Let $x \in X$ with $Ax = y$.

1) If $(x_k)_{k\in\mathbb{N}}$ is generated by the algorithm (5.12) then

(5.13) $\qquad x_k = x_o + P_{k-1}(A)r_o$

where P_{k-1} is a polynomial of degree at most $k - 1$.

2) If $(x'_k)_{k\in\mathbb{N}}$ is another sequence generated by a method of the form (5.13) with $x'_o = x_o$ then

$$H(x_k) \leq H(x_k')$$

where H is the error functional given by

$$H(u) := (x - u, A(x - u)), \quad u \in X.$$

<u>Proof:</u> To prove 1) it is sufficient to show that

$$(*) \qquad r_k = r_o + AP_{k-1}(A)r_o, \quad k \in \mathbb{N} ,$$

for a polynomial P_{k-1} of degree at most $k - 1$. We prove $(*)$ by induction. The cases $k = 0$ and $k = 1$ are clear:

$$r_o = r_o + A\theta r_o \;,\quad r_1 = r_o + A(- \gamma_o E)r_o.$$

By construction

$$r_{k+1} = r_k - \gamma_k As_k$$

and from 5) in Cor.5.10

$$s_k = \tilde{P}_k(A)r_o$$

where $\tilde{P}_k$ is a polynomial of degree at most k. Using $(*)$ we obtain

$$r_{k+1} = r_o - A(P_{k-1}(A) + \tilde{P}_k(A))r_o$$

and $(*)$ is proved for $k + 1$.

To prove 2) we may assume without loss of generality $x_o = \theta$. A simple calculation shows

$$H(x_k') - H(x_k) \geq 2(y - Ax_k, x_k - x_k') + (x_k - x_k', A(x_k - x_k'))$$

$$\geq 2(y - Ax_k, x_k - x_k').$$

Since

$$x_k - x_k' \in \mathrm{span}\{s_o, \ldots, s_{k-1}\}$$

by 7) in Cor.5.10 and therefore

$$(y - Ax_k, x_k - x_k') = (r_k, x_k - x_k') = 0$$

by 3) in Cor.5.10 the result follows.

$\square$

Now we apply the algorithm (5.8) to solve the equation

$$(A + \alpha E)x = y^\varepsilon.$$

Clearly, there exists for each $\alpha > 0$ a unique vector $x^{\alpha, \varepsilon}$ with

(5.14) $(A + \alpha E)x^{\alpha, \varepsilon} = y^\varepsilon.$

The algorithm (5.12) applied with $y := y^\varepsilon$ generates a sequence $(x^{\alpha, \varepsilon, k})_{k \in \mathbb{N}}$. The next theorem gives an upper bound for

$$\|x^{\alpha, \varepsilon, k} - x^o\|_o$$

under a certain "smoothness" assumption. We formulate this assumption by the Hilbert scale $H_s(A)$ which is generated by the compact operator A (see Section 5.2).

Theorem 5.12

Suppose that $x^o \in H_{1/2}(A)$ and that $x_o = 0$. Then

$$\|x^{\alpha, \varepsilon, k} - x^o\| \leq \frac{\varepsilon}{\alpha} + \frac{\sqrt{\alpha}}{2} \|x^o\|_{1/2} +$$

$$+ \sqrt{\frac{a+\alpha}{\alpha}} \left(\frac{1}{k} \sum_{j=0}^{k-1} \frac{\sigma_i}{\sigma_i + \alpha} \right)^k \left\{ \frac{\varepsilon}{\alpha} + \frac{\sqrt{\alpha}}{2} \|x^o\|_{1/2} + \sqrt{a} \|x^o\|_{1/2} \right\}$$

where $a := \sigma_1$.

Proof: We have

$$\|x^{\alpha, \varepsilon, k} - x^o\| \leq \|x^{\alpha, \varepsilon, k} - x^{\alpha, \varepsilon}\| + \|x^{\alpha, \varepsilon} - x^{\alpha, o}\| + \|x^{\alpha, o} - x^o\|$$

and we give a bound for each term on the right-hand side.

We have

$$Ax^o = y \ , \quad x^o = \sum_{j=1}^{\infty} \sigma_j^{-1} (y^o, e_j) e_j \ ,$$

$$(A + \alpha E) x^{\alpha,o} = y^o, \quad x^{\alpha,o} = \sum_{j=1}^{\infty} \frac{1}{\sigma_j + \alpha} (y^o, e_j) e_j.$$

Therefore by the assumption $x^o \in H_{1/2}(A)$ we obtain

$$\begin{aligned}
\| x^o - x^{\alpha,o} \|^2 &= \sum_{j=1}^{\infty} \left(\frac{1}{\sigma_j} - \frac{1}{\sigma_j + \alpha} \right)^2 |(y^o, e_j)|^2 \\
&= \sum_{j=1}^{\infty} \left(\frac{1}{\sigma_j} - \frac{1}{\sigma_j + \alpha} \right)^2 \sigma_j^2 |(x^o, e_j)|^2 \\
&= \frac{\alpha}{4} \sum_{j=1}^{\infty} \left\{ \frac{\sqrt{\alpha}\sqrt{\sigma_j}}{\frac{\sigma_j + \alpha}{2}} \right\}^2 \sigma_j^{-1} |(x^o, e_j)|^2 \\
&\leq \frac{\alpha}{4} \sum_{j=1}^{\infty} \sigma_j^{-1} |(x^o, e_j)|^2 = \frac{\alpha}{4} \| x^o \|_{1/2}^2 \ .
\end{aligned}$$

(1)

It is easy to see that

$$(2) \qquad \| x^{\alpha,o} - x^{\alpha,\varepsilon} \| \leq \frac{\varepsilon}{\alpha} \ .$$

Let

$$Q_k(\lambda) := \prod_{j=0}^{k-1} \frac{\sigma_j - \lambda}{\sigma_j + \alpha} = \prod_{j=0}^{k-1} \left(1 - \frac{\lambda + \alpha}{\sigma_j + \alpha} \right)$$

$$= 1 + (\lambda + \alpha) \ \tilde{P}_{k-1}(\lambda + \alpha) \ .$$

Then $\tilde{P}_{k-1}$ is a polynomial of degree at most $k - 1$. Using the optimality of the conjugate gradient method we obtain with the sequence

$$x_k' := -(A + \alpha E) \ \tilde{P}_{k-1} \ (A + \alpha E) (x^{\alpha,\varepsilon})$$

the following estimate

$$\begin{aligned}
\alpha \| x^{\alpha,\varepsilon} - x^{\alpha,\varepsilon,k} \|^2 &\leq (x^{\alpha,\varepsilon} - x^{\alpha,\varepsilon,k} \ , (A + \alpha E) (x^{\alpha,\varepsilon} - x^{\alpha,\varepsilon,k})) \\
&\leq (x^{\alpha,\varepsilon} - x_k' , (A + \alpha E) (x^{\alpha,\varepsilon} - x_k'))
\end{aligned}$$

$$\leq (a + \alpha) \, \| x^{\alpha, \varepsilon} - x'_k \|^2 \, .$$

Since

$$x^{\alpha, \varepsilon} - x'_k = (E + (A + \alpha E) \, P_{k-1} \, (A + \alpha E)) (x^{\alpha, \varepsilon})$$

$$= Q_k(A)(x^{\alpha, \varepsilon})$$

$$= \sum_{j=1}^{\infty} Q_k(\sigma_j)(x^{\alpha, \varepsilon}, e_j) e_j$$

$$= \sum_{j=k}^{\infty} \left(\prod_{i=0}^{k-1} \frac{\sigma_i - \sigma_j}{\sigma_i + \alpha} \right) (x^{\alpha, \varepsilon}, e_j) e_j$$

we obtain

$$\| x^{\alpha, \varepsilon} - x'_k \|^2 = \sum_{j=k}^{\infty} \left(\prod_{i=0}^{k-1} \frac{\sigma_i - \sigma_j}{\sigma_i + \alpha} \right)^2 \, | (x^{\alpha, \varepsilon}, e_j) |^2 .$$

Using the inequality between the arithmetic and geometric means we obtain

$$\sup_{j \geq k} \left(\prod_{i=0}^{k-1} \frac{\sigma_i - \sigma_j}{\sigma_i + \alpha} \right)^2 \leq \left(\prod_{i=0}^{k-1} \frac{\sigma_i}{\sigma_i + \alpha} \right)^2 \leq \left(\frac{1}{k} \sum_{i=0}^{k-1} \frac{\sigma_i}{\sigma_i + \alpha} \right)^{2k}$$

and therefore

$$(3) \quad \| x^{\alpha, \varepsilon} - x^{\alpha, \varepsilon, k} \| \leq \sqrt{\frac{a + \alpha}{\alpha}} \left(\frac{1}{k} \sum_{i=0}^{k-1} \frac{\sigma_i}{\sigma_i + \alpha} \right)^k \| x^{\alpha, \varepsilon} \| .$$

Since

$$\| x^{\alpha, \varepsilon} \| \leq \| x^{\alpha, \varepsilon} - x^{\alpha, 0} \| + \| x^{\alpha, 0} - x^0 \| + \| x^0 \|$$

and

$$\| x^0 \|^2 = \sum_{j=1}^{\infty} | (x^0, e_j) |^2 = \sum_{j=1}^{\infty} \sigma_j \sigma_j^{-1} | (x^0, e_j) |^2 \leq a \| x^0 \|_{1/2}^2$$

the result follows from (1) – (3). $\qquad \qquad \square$

Corollary 5.13

Suppose that $x^0 \in H_{1/2}(A)$ and that $x_o = \theta$. Then the conjugate gradient method applied to equation (5.10) with $\alpha = \varepsilon^{2/3}$

generates a sequence $(x^{\alpha,\varepsilon,k})_{k\in\mathbb{N}}$ such that

$$\| x^\circ - x^{\alpha,\varepsilon,k(\varepsilon)} \| = O(\varepsilon^{1/3})$$

if we choose the stopping index $k(\varepsilon)$ in an appropriate way.

<u>Proof:</u> This follows from Th.5.12 by choosing $k(\varepsilon)$ such that

$$c_k := \left(\frac{1}{k} \sum_{j=0}^{k-1} \frac{\sigma_j}{\sigma_j+\alpha}\right)^k \leq \varepsilon^{1/3}(a+\varepsilon^{2/3})^{-1}\{\varepsilon^{1/3} + \tfrac{1}{2}\varepsilon^{1/3}\|x^\circ\|_{1/2} + \sqrt{a}\|x^\circ\|_{1/2}\}$$

(Notice that $\lim\limits_{k} c_k = 0$). $\qquad\qquad\square$

5.4 n-widths

Let H be a Hilbert space with inner product $(.,.)$. Given a subset K of H, one might ask how well one can approximate the set K by "simple" subsets of H. A measure how well K can be approximated by n-dimensional subspaces of H is the n-width of K in H. It is defined as follows:

Definition 5.14

Let K be a subset in the Hilbert space H. The
<u>n-width of K in H</u> is given by

$$d_n(K;H) := \inf \left\{ \sup_{x\in K}\ \inf_{y\in U}\|x-y\| \ \middle| \ \begin{array}{l} U \text{ subspace of } H \\ \dim U \leq n \end{array} \right\}.$$

If $d_n(K;H) = \sup\limits_{x\in K}\ \inf\limits_{y\in U^*}\|x-y\|$ for some subspace U^* of dimension at most n, then U^* is said to be an <u>optimal subspace</u>
<u>for $d_n(K;H)$.</u>

In this section we compute the n-width of a subset K in a simple case.
Let H be a separable Hilbert space with orthonormal basis $(e_j)_{j\in\mathbb{N}}$ and let $(\alpha_j)_{j\in\mathbb{N}}$ be a sequence with

$$0 < \alpha_{j+1} \leq \alpha_j\ ,\quad j \in \mathbb{N}\ ,\quad \lim_{j} \alpha_j = 0\ .$$

Then we know from Section 5.1 that there exists a scale $(H_s)_{s \in \mathbb{R}}$ of Hilbert spaces with the properties listed in Th.5.3. Let

$$K := \{x \in H_1 \mid \|x\|_1 \leq 1\}.$$

Theorem 5.15

We have $d_n(K;H) = \alpha_{n+1}$ and $U^* := \langle e_1, \ldots, e_n \rangle$ is an optimal subspace for $d_n(K;H)$.

Proof: We have

$$K = \{x \in H \mid \sum_{j=1}^{\infty} \alpha_j^{-2} |(x,e_j)|^2 \leq 1\}.$$

Let U be a subspace of H with $\dim U \leq n$. Then there exists $\overline{x} \in \langle e_1, \ldots, e_{n+1} \rangle \cap U^{\perp}$ with $\|\overline{x}\|_1 = 1$. If $y \in U$ we have

$$\|\overline{x} - y\|^2 = \|\overline{x}\|^2 - 2(\overline{x},y) + \|y\|^2 \geq \|\overline{x}\|^2 \geq \alpha_{n+1}^2 \|\overline{x}\|_1^2 = \alpha_{n+1}^2$$

and therefore

$$\sup_{x \in K} \inf_{y \in U} \|x - y\| \geq \alpha_{n+1}.$$

Let $x \in K$. Then it is well-known that $\overline{y} := \sum_{j=1}^{n} (x,e_j)e_j$ satisfies

$$\|x - \overline{y}\| = \inf_{y \in U^*} \|x - y\|.$$

Therefore

$$(\inf_{y \in U^*} \|x - y\|)^2 = \sum_{j=n+1}^{\infty} |(x,e_j)|^2$$

$$= \sum_{j=n+1}^{\infty} \alpha_j^2 \alpha_j^{-2} |(x,e_j)|^2 \leq \alpha_{n+1}^2 \|x\|_1^2 \leq \alpha_{n+1}^2$$

and

$$\sup_{x \in K} \inf_{y \in U^*} \|x - y\| \leq \alpha_{n+1}$$

$\square$

Remark 5.16
____

The quantity $e(\varepsilon) := \sup \{|(u,x)| \mid \|Bx\| \le 1, \|Ax\| \le \varepsilon\}$ considered in Section 3.5(which was devoted to the optimal reconstruction of linear functionals)is $d_o(K;\mathbb{R})$ for

$$K := \{(u,x) \mid \|Bx\| \le 1, \|Ax\| \le \varepsilon\}.$$

Bibliographical comments
____

The most complete treatment of Hilbert scales is KREIN, PETUNIN [64]. The convergence of regularizing schemes is discussed in GROETSCH [45], ENGL [34] and HOFMANN [55]. A classical reference concerning the conjugate gradient method for ill-posed equations is KAMMERER, NASHED [62]; a detailed discussion can be found in [52]. The theory of n-widths is the subject of a book by PINKUS [82].

Chapter 6: The method of Tikhonov

In this chapter we discuss the best known regularization method to solve ill-posed problems. We also consider the question how to discretize the problem in an optimal way.

6.1 The generalized inverse

Throughout this chapter, suppose that X and Y are Hilbert spaces and let $A \in B(X,Y)$. We consider the problem of solving the linear equation

$$(6.1) \qquad Ax = y$$

where $y \in Y$ is given. If the equation has no solution (in the classical sense) we may try to find a vector $\bar{x} \in X$ which "comes closest" to solving (6.1) in the sense that

$$\|A\bar{x} - y\| \leq \|Au - y\|$$

for all $u \in X$. To analyse this type of solution is the content of this section.

Let Q denote the projection of Y onto $\overline{R(A)}$, that is

$$\|Qy - y\| \leq \|z - y\| \text{ for all } z \in \overline{R(A)} \text{ , } Qy \in \overline{R(A)} \text{ } (y \in Y).$$

From the projection theorem follows that for all $y \in Y$

$$(6.2) \qquad y - Qy \in \overline{R(A)}^{\perp} = R(A)^{\perp}.$$

Theorem 6.1

Let $y \in Y$. Then the following conditions on $x \in X$ are equivalent:

a) $Ax = Qy$.

b) $\|Ax - y\| \leq \|Au - y\|$ for any $u \in X$.

c) $A^{*}Ax = A^{*}y$.

<u>Proof:</u>

a) $\Longrightarrow$ b) Let $u \in X$. Since $Qy - y \in \overline{R(A)}^{\perp} = R(Q)^{\perp}$ we obtain

$$\|Au - y\|^2 = |Au - Qy|^2 + \|Qy - y\|^2$$

$$= \|Au - Qy\|^2 + \|Ax - y\|^2 \geq \|Ax - y\|^2.$$

b) $\Longrightarrow$ c) Since $Qy \in \overline{R(A)}$ there exists a sequence $(x^n)_{n \in \mathbb{N}}$ in X with $Qy = \lim_n Ax^n$. Then

$$\|Qy - y\|^2 = \lim_n \|Ax^n - y\|^2 \geq \|Ax - y\|^2 \, ,$$

$$|Ax - y|^2 = \|Ax - Qy\|^2 + \|Qy - y\|^2$$

$$\geq \|Ax - Qy\|^2 + \|Ax - y\|^2.$$

This gives

$$Ax = Qy, \quad Ax - y = Qy - y \in \overline{R(A)}^{\perp} = N(A^*),$$

and therefore

$$A^*(Ax - y) = \Theta.$$

c) $\Longrightarrow$ a) We have $Ax - y \in N(A^*) = \overline{R(A)}^{\perp}$ and therefore

$$\Theta = Q(Ax - y) = QAx - Qy = Ax - Qy.$$

$\square$

<u>Definition 6.2</u>

A vector $x \in X$ which satisfies the equivalent conditions a) - c) of Th. 6.1 is called a <u>least squares solution</u> of the equation $Ax = y$.

<u>Corollary 6.3</u>

Let $y \in Y$. Then:

1) $L(y) := \{x \in X \mid A^*Ax = A^*y\} \neq \emptyset$ if and only if $y \in R(A) \oplus R(A)^{\perp}$.

2) If $y \in R(A) \oplus R(A)^{\perp}$ then $L(y)$ is a nonvoid closed convex subset of X.

Proof:

1) Let $x \in L(y)$. Then we know from Theorem 6.1 that
$$Ax - y \in N(A^*) = \overline{R(A)}^{\perp} = R(A)^{\perp}.$$

Therefore $y = Ax + (y - Ax) \in R(A) \oplus R(A)^{\perp}$.

Let $y = y^1 + y^2$ with $y^1 \in R(A)$, $y^2 \in R(A)^{\perp}$. Then
$$Qy = y^1, \quad y^1 = Ax \quad \text{for some } x \in X.$$

Theorem 6.1 shows $x \in L(y)$.

2) This follows immediately from 1) using condition c) in Theorem 6.1. $\square$

By Cor.6.3 and the projection theorem the set $L(y)$ of least squares solutions of the equation $Ax = y$ has an element x of minimal norm if $y \in R(A) \oplus R(A)^{\perp}$:
$$\|x\| = \min\{\|u\| \mid u \in L(y)\}.$$

Therefore the following definition makes sense.

Definition 6.4

The mapping A^+ with domain of definition
$$D(A^+) := R(A) \oplus R(A)^{\perp}$$
defined for $y \in D(A^+)$ by $A^+ y := x$ where x is the least squares solution of minimal norm of the equation $Ax = y$ is called the generalized inverse.

Corollary 6.5

We have:

1) $D(A^+)$ is dense in Y and $D(A^+) = Y$ if $R(A)$ is closed.

94

2) If $R(A)$ is closed and A^{-1} exists then $A^+\big|_{R(A)} = A^{-1}$.

3) $R(A^+) = N(A)^{\perp}$.

4) A^+ is a linear mapping.

5) A^+ is bounded ($\sup\{\|A^+y\|\,\big|\,y \in D(A^+),\ \|y\| \leq 1\} < \infty$)
   if and only if $R(A)$ is closed.

6) For each $y \in D(A^+)$ the vector A^+y is the unique least
   squares solution of $Ax = y$ lying in $N(A)^{\perp}$.

<u>Proof:</u>

1) $\overline{D(A^+)} = \overline{R(A) \oplus R(A)^{\perp}} \supset \overline{R(A)} \oplus \overline{R(A)}^{\perp} = Y$.
   This shows that $D(A^+)$ is dense in Y. If $R(A)$ is closed then
   $D(A^+) = R(A) \oplus R(A)^{\perp} = \overline{R(A)} \oplus \overline{R(A)}^{\perp} = Y$.

2) This is an immediate consequence of the definition of A^+.

3) Let $w \in R(A^+)$, $w = A^+y$ with $y \in D(A^+)$. We can decompose w
   into
   $$w = w^1 + w^2,\quad w^1 \in N(A)^{\perp},\ w^2 \in N(A),$$
   and obtain with Theorem 6.1
   $$Aw = A(w^1 + w^2) = Aw^1 = AA^+y = Qy.$$
   This implies that w^1 is a least squares solution of $Ax = y$
   and therefore
   $$\|A^+y\|^2 = \|w\|^2 = \|w^1\|^2 + \|w^2\|^2 \geq \|A^+y\|^2 + \|w^2\|^2.$$
   This implies $w^2 = \theta$, that is $w \in N(A)^{\perp}$.
   Let $w \in N(A)^{\perp}$ and let $y := Aw$. Then
   $$Aw = QAw = Qy$$
   which shows that w is a least squares solution of $Ax = y$.
   Let v be another least squares solution. Then
   $$Av = Qy = Aw,\quad v - w \in N(A),$$
   and
   $$\|v\|^2 = \|w\|^2 + \|v - w\|^2 \geq \|w\|^2.$$
   This implies that w is a least squares solution with mini-
   mal norm, i.e. $w = A^+y$.

4) Let $y, \tilde{y} \in D(A^+)$. Then

$$AA^+ y = Qy, \quad AA^+ \tilde{y} = Q\tilde{y} \; ,$$

$$AA^+ y + AA^+ \tilde{y} = Qy + Q\tilde{y} = Q(y + \tilde{y}) = AA^+(y + \tilde{y}) ,$$

and therefore by 3)

$$A^+ y + A^+ \tilde{y} - A^+(y + \tilde{y}) \in N(A) \cap N(A)^\perp .$$

This implies

$$A^+(y + \tilde{y}) = A^+ y + A^+ \tilde{y} \; .$$

In a similar way it can be shown that for any scalar α

$$A^+(\alpha y) = \alpha A^+ y .$$

5) Let A^+ be bounded. Since

$$AA^+ y = Qy \quad \text{for all} \quad y \in D(A^+)$$

and $D(A^+)$ is dense in Y the mapping A^+ may be extended to
a bounded operator $\tilde{A} \in B(Y,X)$ with

$$A\tilde{A} y = Qy \quad \text{for all} \quad y \in Y .$$

This shows

$$\overline{R(A)} = R(Q) \subset R(A)$$

from which we conclude that $R(A)$ is closed.

Now suppose that $W := R(A)$ is closed. Since $R(A^+) = N(A)^\perp$
and $X = N(A) \oplus N(A)^\perp$ the mapping

$$\hat{A} : N(A)^\perp \longrightarrow W , \quad \hat{A} u := Au ,$$

is bijective. Since $\hat{A}$ is continuous and W is a Hilbert
space as a closed subspace of Y the inverse $\hat{A}^{-1}$ exists and
is continuous by Banach's theorem. Therefore there exists
$m > o$ with

$$\|A^+ y\| = \|\hat{A}^{-1}(A\hat{A}^+ y)\| \leq m \|\hat{A}A^+ y\| = m \|AA^+ y\|$$

for all $y \in D(A^+) = Y$ (see 1)). From this we obtain

$$\|y\| \geq \|Qy\| = \|AA^+ y\| \geq m^{-1}\|A^+ y\| , \quad y \in Y .$$

This shows $A^+ \in B(Y,X) , \quad \|A^+\| \leq m$.

6) Follows from the definition of $A^+ y$ and the fact that
$R(A^+) = N(A)^\perp$.

If the operator A is compact we can give an explicit representation of A^+ by using a singular system of A.

Theorem 6.6

Let $A : X \longrightarrow Y$ be compact and let $(\sigma_j, e_j, f_j)_{j \in \mathbb{N}}$ be a singular system of A. Then

$$A^+ y = \sum_{j=1}^{\infty} \sigma_j^{-1} (Qy, f_j) e_j = \sum_{j=1}^{\infty} \sigma_j^{-1} (y, f_j) e_j, \quad y \in D(A^+).$$

Proof: We note that $Qy \in R(A)$ since $y \in D(A^+)$. Therefore by Th.4.22

$$\sum_{j=1}^{\infty} \sigma_j^{-2} |(Qy, f_j)|^2 < \infty.$$

Also note that since $f_j \in \overline{R(A)}$ for all $j \in \mathbb{N}$

$$(Qy, f_j) = (y, Qf_j) = (y, f_j), \quad j \in \mathbb{N}.$$

We therefore see that the series

$$\sum_{j=1}^{\infty} \sigma_j^{-1} (Qy, f_j) e_j, \quad \sum_{j=1}^{\infty} \sigma_j^{-1} (y, f_j) e_j$$

converge. Also, since each e_j lies in $N(A)^{\perp}$, it follows that the vector

$$x := \sum_{j=1}^{\infty} \sigma_j^{-1} (Qy, f_j) e_j$$

is also in $N(A)^{\perp}$. Moreover,

$$Ax = \sum_{j=1}^{\infty} \sigma_j^{-1} (Qy, f_j) Ae_j = \sum_{j=1}^{\infty} (Qy, f_j) f_j = Qy$$

since $Qy \in R(A)$ and $\text{span}\{f_j | j \in \mathbb{N}\} = \overline{R(A)}$ (see Th.4.14). Therefore x is a least squares solution of $Ax = y$ in $N(A)^{\perp}$, i.e., $x = A^+ y$.

$\square$

At first sight it seems that the most natural way to compute least squares solutions is to use the condition c) in Th. 6.1. However, even if for a given y the equation

$$A^*Ax = A^+y$$

has a unique solution x, a small perturbation of y may result in a drastic change in x since A^+ is generally discontinuous. Since in practical problems the right-hand side y of the equation Ax = y is usually the result of measurements and therefore the assumption $y \in D(A^+)$ is not satisfied the solution-concept "generalized inverse" is not applicable. To overcome this difficulty we might try to replace the operator A^*A in the equation $A^*Ax = A^*y$ by some nearby operator which has a bounded inverse. With the realization of such methods we are concerned in the next section.

6.2 The classical method of Tikhonov

Let us consider the following problem:

$$(6.3) \qquad \text{Minimize } \|Ax - y\|^2 + \alpha\|x\|^2 \text{ in } X \ (\alpha > o, y \in Y).$$

This problem is a special case of the minimization problem investigated in Section 3.2. From Theorem 3.7 we obtain

Theorem 6.7

The problem (6.3) has for each $\alpha > o$ and $y \in Y$ a unique solution x^α and this solution is characterized as the unique element x^α in X satisfying

$$(6.4) \qquad (A^*A + \alpha I)x^\alpha = A^*y.$$

The solution x^α of (6.3) depends continuously on y. This follows from the inequality

$$\|(A^*A + \alpha I)x\| \geq \alpha\|x\| \ , \ x \in X.$$

Next, we want to show that $(x^\alpha)_{\alpha > o}$ converges to A^+y as $\alpha \longrightarrow o$ for each $y \in D(A^+)$.

Theorem 6.8

Let $y \in D(A^+)$ and define

$$\delta_R := \inf \{\|A^+y - A^*w\| \mid \|w\| \le R\} \, , \quad R > 0.$$

Then

1) $\lim\limits_{R \to 0} \delta_R = 0$.

2) $\|A^+y - x^\alpha\| \le (\delta_R^2 + \alpha R^2)^{1/2}$ for all $\alpha > 0$ and $R > 0$.

3) $\lim\limits_{\alpha \to 0} x^\alpha = A^+y$.

<u>Proof:</u> Since $R(A^+) = N(A)^\perp = \overline{R(A^*)}$ the result 1) is immediately clear.

We have with $x := A^+y$

$$(A^*A + \alpha I)x^\alpha = A^*y, \quad A^*Ax = A^*y.$$

Let $w \in Y$ with $\|w\| \le R$. Then

$$\|Ax^\alpha - Ax\|^2 = (Ax^\alpha - Ax, Ax^\alpha - Ax)$$

$$= -\alpha\|x^\alpha - x\|^2 - \alpha(x, x^\alpha - x)$$

$$= -\alpha\|x^\alpha - x\|^2 - \alpha(x - A^*w, x^\alpha - x) - \alpha(w, A(x^\alpha - x))$$

$$\le -\alpha\|x^\alpha - x\|^2 + \alpha\|x - A^*w\|\,\|x^\alpha - x\| + \alpha\|w\|\,\|Ax^\alpha - Ax\|$$

$$\le -\alpha\|x^\alpha - x\|^2 + \alpha\delta_R\|x^\alpha - x\| + \alpha R\|Ax^\alpha - Ax\|$$

$$\le -\alpha\|x^\alpha - x\|^2 + \alpha\{\tfrac{1}{2}\delta_R^2 + \tfrac{1}{2}\|x^\alpha - x\|^2\} + \tfrac{1}{2}\alpha^2 R^2 + \tfrac{1}{2}\|Ax^\alpha - Ax\|^2.$$

This gives

$$\tfrac{\alpha}{2}\|x^\alpha - x\|^2 + \tfrac{1}{2}\|Ax^\alpha - Ax\|^2 \le \tfrac{\alpha}{2}\delta_R^2 + \tfrac{\alpha^2}{2}R^2$$

and we obtain the result 2).

Using a simple limit argument we get from 2) the result 3). $\quad\square$

<u>Example 6.9</u>

If $x = A^+y \in R(A^*)$, so $x = A^*w$ for some $w \in Y$, we have $\delta_R = 0$ for $R := \|w\|$ and the estimate in 2) of Th.6.8 simplifies to

$$\|x^\alpha - x\| \le \sqrt{\alpha}\,\|w\|. \qquad\qquad *$$

If the operator A is compact with singular system $(\sigma_j, e_j, f_j)_{j \in \mathbb{N}}$ then from the equation (6.4) we get that x may be represented by

$$x^\alpha = \sum_{j=1}^\infty q(\alpha, \sigma_j) \sigma_j^{-1} (y, f_j) e_j$$

where $q(\alpha, \sigma) := \dfrac{\sigma^2}{\sigma^2 + \alpha}$. This type of regularization was already considered in Section 5.2.

6.3 Error bounds for Tikhonov regularization in Hilbert scales.

Let $x^o \in X$ and $y^o, y^\varepsilon \in Y$ with

$$(6.5) \qquad Ax^o = y^o, \quad \|y^o - y^\varepsilon\| \leq \varepsilon$$

where $\varepsilon \geq o$ is a known noise-level. If our a-priori-information about x^o is given by

$$(6.6) \qquad x^o \in H_q, \quad \|x^o\|_q \leq E \qquad (q \geq o, E \geq o),$$

where $(H_s)_{s \in \mathbb{R}}$ is a Hilbert scale constructed by an orthonormal basis $(e_j)_{j \in \mathbb{N}}$ in $H_o := X$ and a sequence $(\alpha_j)_{j \in \mathbb{N}}$ (see Section 5.1) then the following variant of Tikhonov regularization can be considered:

$$(6.7) \qquad \text{Minimize } \|Ax - y^\varepsilon\|^2 + \frac{\varepsilon^2}{E^2} \|x\|_q^2 \text{ in } H_q .$$

The minimizer in (6.7) is called the Tikhonov regularized solution. From Section 3.2 it is clear that there exists a unique minimizer x^ε in (6.7) if $\varepsilon > o$. Moreover, we know from Th.3.4 that the regularization error $x^\varepsilon - x^o$ can be estimated as follows:

$$(6.8) \qquad \|x^\varepsilon - x^o\|_o \leq 2\sqrt{2}\, \omega(\varepsilon, E)$$

where

$$\omega(\varepsilon, E) := \sup\{\|x\|_o \,\big|\, x \in H_q, \|x\|_q \leq E, \|Ax\| \leq \varepsilon\}.$$

In order to compute $\omega(\varepsilon, E)$ we need an information on the "degree of ill-posedness":

100

There exists $a > 0$ and $M \geq m > 0$ such that

$$\text{(6.9)} \qquad m\|x\|_{-a} \leq \|Ax\| \leq M\|x\|_{-a} \quad \text{for all } x \in H_o .$$

Then from the interpolation inequality (see Theorem 5.3)

$$\|x\|_o \leq \|x\|_{-a}^{\frac{q}{q+a}} \; \|x\|_q^{\frac{a}{q+a}} \qquad , \quad x \in H_q \; ,$$

follows immediately the estimate

$$\text{(6.10)} \qquad \omega(\varepsilon,E) \leq m^{-\frac{q}{q+a}} \; \varepsilon^{\frac{a}{q+a}} \; E^{\frac{a}{q+a}}$$

so that the following theorem holds:

Theorem 6.10

If the assumptions (6.6) and (6.9) are satisfied then we have

$$\|x^\varepsilon - x^o\|_o \leq S \, \varepsilon^{\frac{q}{q+a}} \; E^{\frac{a}{q+a}}$$

where S is a constant independent of ε and E.

Instead of (6.7) one may consider the following variant of Tikhonov regularization:

$$\text{(6.11)} \qquad \text{Minimize } \|Ax - y^\varepsilon\|^2 + \alpha\|x\|_p^2 \text{ in } H_p .$$

The difference to the method (6.7) consists in the fact that the a-priori-information is given in H_q (see (6.6)) and the regularization is formulated with H_p.

Let $x^{\varepsilon,\alpha,p}$ be the minimizer in (6.11) which exists if $\alpha > 0$ and $p \geq 0$. We want to find out for which values of p the error $\|x^{\varepsilon,\alpha,p} - x^o\|_o$ has the same order as $\omega(\varepsilon,E)$ provided α is chosen properly. The analysis above shows that the choice $p = q$, $\alpha = \dfrac{\varepsilon^2}{E^2}$ leads to the correct order. We analyse this question only in a very simple situation, namely that A and $(H_s)_{s \in \mathbb{R}}$ are given as follows:

$$\text{i)} \quad A : H_o \longrightarrow H_o \text{ is an injective compact operator}$$

(6.12) with singular system $(\sigma_j, e_j, f_j)_{j \in \mathbb{N}}$;

$$\text{ii)} \quad H_s = H_s(A), \; s \in \mathbb{R} \; \text{(see Section 5.1).}$$

In this situation the assumption (6.9) is satisfied with $a = 1$ and we have formally

$$x^o = \sum_{j=1}^{\infty} (x^o, e_j) e_j \; ,$$

$$x^{\varepsilon, \alpha, p} = \sum_{j=1}^{\infty} (\sigma_j^2 + \alpha \sigma_j^{-2p})^{-1} \sigma_j (y^{\varepsilon}, f_j) e_j \; ,$$

$$x^{o, \alpha, p} = \sum_{j=1}^{\infty} (\sigma_j^2 + \alpha \sigma_j^{-2p})^{-1} \sigma_j^2 (x^o, e_j) e_j .$$

If we assume

$$(6.13) \qquad p \geq \frac{1}{2} q - 1$$

we obtain

$$\| x^{\varepsilon, \alpha, p} - x^{o, \alpha, p} \|_o \leq c_1 \varepsilon \alpha^{-\frac{1}{2p+2}} \; ,$$

$$\| x^{o, \alpha, p} - x^o \|_o \leq c_2 \, E \alpha^{\frac{q}{2p+2}} \; ,$$

where c_1, c_2 are independent of α and ε. This implies

$$(6.14) \quad \| x^{\varepsilon, \alpha, p} - x^o \|_o \leq S \varepsilon^{\frac{q}{q+1}} E^{\frac{1}{q+1}}$$

if we choose

$$(6.15) \qquad \alpha = O(\varepsilon^{2\frac{p+1}{q+1}} E^{-2\frac{p+1}{q+1}}) ;$$

here S is a constant independent of ε and E.

This shows that the method (6.11) under the assumption (6.6) and (6.13) in the situation (6.12) leads to an optimal estimate if we use the parameter choice strategy (6.15). Notice that the assumption (6.13) is sufficient to ensure that

$$x^{\varepsilon, \alpha, p} \in H_q .$$

6.4. On discrepancy principles

In Section 6.3 we obtained optimal convergence results for Tikhonov regularization under suitable assumptions using the a-priori parameter strategy

$$\alpha = O\left(\frac{\varepsilon^2}{E^2}\right) \qquad\qquad (\text{Method}) \quad (6.7)$$

and

$$\alpha = O\left(\varepsilon^{2\,\frac{p+1}{q+1}}\, E^{-2\,\frac{p+1}{q+1}}\right) \qquad (\text{Method}) \quad (6.11)$$

In contrast to the a-priori choice of α one may try to find a criterion for the choice of α where results occuring during the computations are used. Such strategies are called a-posteriori strategies.

A widely used a-posteriori choice of α is the so-called discrepancy principle where α is chosen as the (unique) solution in the equation

$$(6.16) \qquad \|Ax^{\alpha,\varepsilon} - y^\varepsilon\|^2 = \varepsilon^2 \;;$$

here $x^{\alpha,\varepsilon}$ is Tikhonov's regularized solution given as the solution of the problem

$$\text{Minimize } \|Ax - y^\varepsilon\|^2 + \alpha\|x\|_q^2 \text{ in } H_q.$$

Another method is to choose α as the (unique) solution of

$$(6.17) \qquad \|Ax^{\alpha,\varepsilon} - y^\varepsilon\|^2 = \varepsilon^2\,\alpha^{-1}.$$

One can show that neither the Method(6.16)nor the Method(6.17) yield the optimal convergence rate. In this section we consider a variant of the discrepancy principle which gives rise to optimal convergence rates.

With the family $(x^{\alpha,\varepsilon})_{\alpha>0}$ we define

$$\rho(\alpha,\varepsilon) := \alpha^2\|x^{\alpha,\varepsilon}\|_q^2$$

and the criterion for chosing α consists in the solution of the equation

$$(6.18) \qquad \rho(\alpha,\varepsilon) = \varepsilon^\nu\alpha^{-\mu} \qquad (\mu \geq 0,\ \nu \geq 0).$$

Let $A^{\#}$ denote the adjoint of $A\big|_{H_q}$ with respect to the Hilbert scale $(H_s)_{s\in\mathbb{R}}$:

$$A^{\#} : Y \longrightarrow H_{-q}$$

$$(A^{\#}y,u) = (y,Au) \quad \text{for all } y \in Y,\ u \in H_q.$$

In the following, D_q is the map associated with the Hilbert scale $(H_s)_{s\in\mathbb{R}}$ (see Section 5.1).

<u>Lemma 6.11</u>

$A^{\#}A + \alpha D_{-q}^{-1} D_q$ is for each $\alpha > 0$ a bijective bounded operator from H_q onto H_{-q} with

$$(6.19) \qquad \| (A^{\#}A + \alpha D_{-q}^{-1} D_q)x \|_{-q} \geq \alpha \|x\|_q , \quad x \in H_q.$$

<u>Proof:</u> Let $x \in H_q$, $x \neq 0$. Then

$$\| (A^{\#}A + \alpha D_{-q}^{-1} D_q)x \|_{-q} = \sup\{ \| (A^{\#}A + \alpha D_{-q}^{-1} D_q)x,u)_o \| \mid u \in H_q, \|u\|_q \leq 1\}$$

$$\geq ((A^{\#}A + \alpha D_{-q}^{-1} Dq)x,x)_o \|x\|_q^{-1}$$

$$\geq \alpha (D_{-q}^{-1} D_q x,x)_o \|x\|_q^{-1}$$

$$= \alpha \|x\|_q$$

since $(D_{-q}^{-1} D_q x,x)_o = (D_q x, D_q x)_o = \|x\|_q^2$. This proves the inequality (6.19). From this inequality we conclude that $A^{\#}A + D_{-q}^{-1}D_q$:

$H_q \longrightarrow H_{-q}$ is bounded and injective. Moreover, $R(A^{\#}A + \alpha D_{-q}^{-1} D_q)$ is closed due to (6.19). Therefore $R(A^{\#}A + \alpha D_{-q}^{-1} D_q) = H_{-q}$ by a simple application of the projection theorem. $\qquad\qquad\Box$

Let $A^* := D_q^{-1} D_{-q}^{-1} A^{\#}$. Then

$$A^* : Y \longrightarrow H_q, \quad (A^*y,u)_q = (y,Au) \quad \text{for all } y \in Y,\ u \in H_q.$$

In the following we always assume

(6.20) $A^{\#}y^{\varepsilon} \neq \theta$ $(A^{*}y^{\varepsilon} \neq \theta)$

an assumption which excludes only trivial cases.

Lemma 6.12

We have

1) $\rho(\alpha,\varepsilon) = \|A^{\dagger}Ax^{\alpha,\varepsilon} - A^{\#}y^{\varepsilon}\|_{-q}^{2} = \|A^{*}Ax^{\alpha,\varepsilon} - A^{*}y^{\varepsilon}\|_{q}^{2}$, $\alpha > o$.

2) For each $\mu > o$ and $\nu \geq o$ there exists a unique α with

$$\rho(\alpha,\varepsilon) = \varepsilon^{\nu}\alpha^{-\mu}.$$

Proof:

Part 1) follows from

$$A^{\dagger}Ax^{\alpha,\varepsilon} - A^{\#}y^{\varepsilon} = -\alpha\ D_{-q}^{-1}\ D_{q}x^{\alpha,\varepsilon}$$

and

$$\|x^{\alpha,\varepsilon}\|_{q} = \|D_{q}x^{\alpha,\varepsilon}\|_{o} = \|D_{-q}^{-1}\ D_{q}x^{\alpha,\varepsilon}\|_{-1}.$$

To prove part 2) we establish first the following properties:

i) $\rho(.,\varepsilon)$ is continuous .

ii) $\lim\limits_{\alpha \to o} \rho(\alpha,\varepsilon) = O$, $\lim\limits_{\alpha \to \infty} \rho(\alpha,\varepsilon) = \|A^{\#}y^{\varepsilon}\|^{2}$.

iii) $\rho(.,\varepsilon)$ is differentiable and $\frac{\partial\rho}{\partial\alpha}(\alpha,\varepsilon) \geq O$ for all $\alpha > o$.

Clearly, i) follows from iii). Since

$$\|x^{\alpha,\varepsilon}\|_{q}^{2} \leq \frac{1}{\alpha}\ (\alpha\|x^{\alpha,\varepsilon}\|_{q}^{2} + \|Ax^{\alpha,\varepsilon} - y^{\varepsilon}\|^{2})$$

$$\leq \frac{1}{\alpha}\ (\alpha E^{2} + \varepsilon^{2}) = E^{2} + \frac{\varepsilon^{2}}{\alpha}$$

the properties in ii) follow from the identity

$$\rho(\alpha,\varepsilon) = \alpha^{2}\|x^{\alpha,\varepsilon}\|_{q}^{2} = \|A^{\dagger}Ax^{\alpha,\varepsilon} - A^{\#}y^{\varepsilon}\|^{2}.$$

Let $v^{\alpha,\varepsilon} := - (A^{\dagger}A + \alpha D_{-q}^{-1}\ D_{q})^{-1}\ D_{-q}^{-1}\ D_{q}x^{\alpha,\varepsilon}$. Then one can see

that $\frac{\partial\rho}{\partial\alpha}(\alpha,\varepsilon)$ exists and may be represented as

$$\frac{\partial \rho}{\partial \alpha}(\alpha,\varepsilon) = 2\alpha \|x^{\alpha,\varepsilon}\|_q^2 + 2\alpha^2 (x^{\alpha,\varepsilon}, v^{\alpha,\varepsilon})_q .$$

From this we obtain with Lemma 6.11

$$\frac{\partial \rho}{\partial \alpha}(\alpha,\varepsilon) = 2\alpha \|x^{\alpha,\varepsilon}\|_q^2 + 2\alpha^2 (x^{\alpha,\varepsilon}, v^{\alpha,\varepsilon})_q$$

$$= 2\alpha \{ \|x^{\alpha,\varepsilon}\|_q^2 - \alpha (x^{\alpha,\varepsilon}, (A^{\#}A + \alpha D_{-q}^{-1} D_q)^{-1} D_{-q}^{-1} D_q x^{\alpha,\varepsilon})_q \}$$

$$\geq 2\alpha \{ \|x^{\alpha,\varepsilon}\|_q^2 - \alpha \alpha^{-1} \|D_{-q}^{-1} Dq\, x^{\alpha,\varepsilon}\|_{-q}^2 \} \geq 0.$$

Let $f(\alpha) := \alpha^{\mu} \rho(\alpha,\varepsilon)$. Then by iii) we have

$$\frac{df}{d\alpha}(\alpha) = \mu \alpha^{\mu-1} \rho(\alpha,\varepsilon) + \alpha^{\mu} \frac{\partial \rho}{\partial \alpha}(\alpha,\varepsilon) > 0 \text{ for all } \alpha > 0,$$

and

$$\lim_{\alpha \to 0} f(\alpha) = 0 \quad , \quad \lim_{\alpha \to \infty} f(\alpha) = \infty.$$

Thus, it follows that f is continuous, strictly increasing and goes to $+\infty$ as $\alpha \longrightarrow \infty$ and to 0 as $\alpha \longrightarrow 0$. Hence part 2) follows from the mean value theorem. $\quad\square$

In the following, let $\alpha(\varepsilon)$ denote the unique solution of the equation (6.18).

Lemma 6.13

We have

1) $\displaystyle \lim_{\varepsilon \to 0} \alpha(\varepsilon) = 0$,

2) $\displaystyle \lim_{\varepsilon \to 0} \rho(\alpha(\varepsilon),\varepsilon) = 0.$

Proof: Assume that there is a sequence $(\varepsilon_n)_{n \in \mathbb{N}}$ with

$\displaystyle \lim_n \varepsilon_n = 0$ but $\displaystyle \lim_n \alpha(\varepsilon_n) = \infty.$ Since

$$\|x^{\alpha(\varepsilon_n),\varepsilon_n}\|_q = \|(A^{\#}A + \alpha(\varepsilon_n)D_{-q}^{-1}D_q)^{-1}A^{\#}y^{\varepsilon_n}\|_q$$

$$\leq \alpha(\varepsilon_n)^{-1}(\|A^{\#}(y^{\varepsilon_n} - y^o)\|_{-q} + \|A^{\#}y^o\|_{-q})$$

$$\leq \alpha(\varepsilon_n)^{-1}\|A^{\#}\|(\varepsilon_n + \|y^o\|) \quad , \ n \in \mathbb{N} \ ,$$

we have $\lim_n x^{\alpha(\varepsilon_n)\varepsilon_n} = \Theta$. Then $\lim_n \rho(\alpha(\varepsilon_n),\varepsilon_n) = \|A^{\#}y^o\|_{-q}^2$

and therefore

$$O = \lim_n \varepsilon_n^{\nu} = \lim_n(\alpha(\varepsilon_n)^{\mu}\rho(\alpha(\varepsilon_n),\varepsilon_n)) = \infty.$$

This is a contradiction and therefore $\alpha(\varepsilon)$ remains bounded as $\varepsilon \longrightarrow o$. Now assume that there exists a sequence $(\varepsilon_n)_{n\in\mathbb{N}}$ with $\lim_n \varepsilon_n = o$ but $\lim_n \alpha(\varepsilon_n) = \bar{\alpha} > o$. Then

$$\lim_n x^{\alpha(\varepsilon_n),\varepsilon_n} = \lim_n (A^{\#}A + \alpha(\varepsilon_n)D_{-q}^{-1}D_q)^{-1}A^{\#}y^{\varepsilon_n}$$

$$= (A^{\#}A + \bar{\alpha}D_{-q}^{-1}D_q)^{-1}A^{\#}y^o$$

and hence

$$\bar{\alpha}^{\mu}\|A^{\#}A(A^{\#}A + \bar{\alpha}D_{-q}^{-1}D_q)^{-1}A^{\#}y^o - A^{\#}y^o\|_{-q}^2$$

$$= \lim_n (\alpha(\varepsilon_n)^{\mu}\rho(\alpha(\varepsilon_n),\varepsilon_n)) = \lim_n \varepsilon_n^{\nu} = O,$$

so that

$$A^{\#}A(A^{\#} + \bar{\alpha}D_{-q}^{-1})^{-1}A^{\#}y^o = A^{\#}y^o \ ,$$

which contradicts the assumption $A^{\#}y^o \neq \Theta$.

Thus, $\lim_{\varepsilon \to o} \alpha(\varepsilon) = O$.

We have

$$0 \leq \rho(\alpha(\varepsilon),\varepsilon) = \|A^{\#}Ax^{\alpha(\varepsilon),\varepsilon} - A^{\#}y^{\varepsilon}\|_{-q}^{2}$$

$$\leq \|A^{\#}\|^{2}\;\|Ax^{\alpha(\varepsilon),\varepsilon} - y^{\varepsilon}\|^{2}$$

$$\leq \|A^{\#}\|^{2}\{\|Ax^{\alpha(\varepsilon),\varepsilon} - y^{\varepsilon}\|^{2} + \alpha(\varepsilon)\|x^{\alpha(\varepsilon),\varepsilon}\|_{q}^{2}\}$$

$$\leq \|A^{\#}\|^{2}\{\|Ax^{o} - y^{\varepsilon}\|^{2} + \alpha(\varepsilon)\,\|x^{o}\|_{q}^{2}\}$$

$$\leq \|A^{\#}\|^{2}\;\{\varepsilon^{2} + \alpha(\varepsilon)E^{2}\}$$

which implies the result 2). □

Lemma 6.14

If $o < \nu \leq 2\mu$ then

$$\lim_{\varepsilon \to o} \varepsilon^{2}\alpha(\varepsilon)^{-1} = 0 .$$

Proof: We have

$$\varepsilon^{\nu}\alpha(\varepsilon)^{-\frac{\nu}{2}} = \varepsilon^{\nu}(\alpha(\varepsilon)^{-\mu})\alpha(\varepsilon)^{\mu-\frac{\nu}{2}}$$

$$= \rho(\alpha(\varepsilon),\varepsilon)\alpha(\varepsilon)^{\mu-\frac{\nu}{2}}$$

which implies with Lemma 6.13

$$\lim_{\varepsilon \to o} \varepsilon^{2}\alpha(\varepsilon)^{-1} = 0.$$

□

Theorem 6.15

Let the assumptions (6.6),(6.8) and (6.20) hold and assume $\nu = 2\mu + 4,\ \mu > o$. Then

$$\|x^{o} - x^{\alpha(\varepsilon),\varepsilon}\|_{o} = 0(\varepsilon^{\frac{q}{q+a}})$$

Proof: We have

$$(*)\qquad\begin{aligned}
\|Ax^{\alpha(\varepsilon),\varepsilon} - y^\varepsilon\|^2 + \alpha(\varepsilon)\|x^{\alpha(\varepsilon),\varepsilon}\|_q^2 &\le \varepsilon^2 + \alpha(\varepsilon)E^2\ ,\\[2mm]
\|x^{\alpha(\varepsilon),\varepsilon}\|_q &\le E + \alpha(\varepsilon)^{-\frac{1}{2}}\varepsilon\ ,\\[2mm]
\|Ax^{\alpha(\varepsilon),\varepsilon} - y^\varepsilon\| &\le (\varepsilon^2 + \alpha(\varepsilon)E^2)^{\frac{1}{2}}\ ,\\[2mm]
\|A(x^{\alpha(\varepsilon),\varepsilon} - x^0)\| &\le 2(\varepsilon^2 + \alpha(\varepsilon)E^2)^{\frac{1}{2}}\ .
\end{aligned}$$

Using assumption (6.9) we obtain by an interpolation inequality

$$\|x^0 - x^{\alpha(\varepsilon),\varepsilon}\| \le m^{-1}\|A(x^0 - x^{\alpha(\varepsilon),\varepsilon})\|^{\frac{q}{q+a}}(E + E + \alpha(\varepsilon)^{-\frac{1}{2}}\varepsilon)^{\frac{a}{q+a}}.$$

This shows that $\lim_{\varepsilon\to 0} x^{\alpha(\varepsilon),\varepsilon} = x^0$ since $\lim_{\varepsilon\to 0}\varepsilon^2\alpha(\varepsilon)^{-1} = 0$ by Lemma 6.14. This implies that

$$\lim_{\varepsilon\to 0}(\varepsilon^\nu\alpha(\varepsilon)^{-\mu-2}) = \lim_{\varepsilon\to 0}(\alpha(\varepsilon)^{-2}\rho(\alpha(\varepsilon),\varepsilon)) = \|x^0\|_q > 0$$

because of (6.20). Then there are constants $c_1,c_2 > 0$ such that for sufficiently small $\varepsilon > 0$

$$c_1 \le \varepsilon^\nu\alpha(\varepsilon)^{-\mu-2} \le c_2.$$

This implies for sufficiently small $\varepsilon > 0$

$$\|A(x^{\alpha(\varepsilon),\varepsilon} - x^0)\| \le c_3(\varepsilon^2 + \varepsilon^{\frac{\nu}{\mu+2}}E^2)^{\frac{1}{2}}$$

with a constant c_3 independent of ε. The interpolation inequality gives the result. $\qquad\square$

The result of Th.6.15 is that the a-posteriori strategy
derived from the criterion (6.18) leads assymptotically to
the correct estimate provided the numbers μ,ν are chosen properly. There remains the question how the nonlinear equation
(6.18) can be solved efficiently. It can be shown that this
can be achieved by Newton's method; see the proof of Lemma 6.12
for a hint how this method can be realized .

6.5 Discretization in Tikhonov's method

In this section we consider the method (6.7) in a discrete version, that is we replace the minimization of

$$\|Ax - y^\varepsilon\|^2 + \frac{\varepsilon^2}{E^2} \|x\|_q^2$$

in H_q by the minimization over a finite-dimensional subspace.

Let $(X_h)_{h>o}$ be a family of subspaces of X such that the following conditions are satisfied (for the meaning of q and a see (6.6) and (6.9)):

> i) $X_h \subset H_q$, $\dim X_h < \infty$.

> ii) There exists a linear map $I_h : H_q \longrightarrow X_h$
>   such that

(6.21)
$$\|u - I_h u\|_{-a} \leq ch^{a+q} \|u\|_q, \quad \|I_h u\|_q \leq c\|u\|_q$$

>   for all $u \in H_q$ where c is a constant independent of h.

Let $x^{\varepsilon,h}$ be the minimizer of

$$\|Ax - y^\varepsilon\|^2 + \frac{\varepsilon^2}{E^2} \|x\|_q^2$$

in X_h.

Theorem 6.16

Let the assumptions (6.6), (6.9) and (6.21) are satisfied. Then if we choose

(6.22)
$$h = O(\varepsilon^{\frac{1}{q+a}})$$

there exists a constant $c \geq o$ independent of ε, h and E such that

(6.23)
$$\|x^o - x^{\varepsilon,h}\|_o \leq c(1 + E^2)^{1/2} \varepsilon^{\frac{q}{q+a}} .$$

Proof: We have

$$\|Ax^{\varepsilon,h} - y^{\varepsilon}\|^2 + \frac{\varepsilon^2}{E^2} \|x^{\varepsilon,h}\|_q^2 \leq \|AI_h x^o - y^{\varepsilon}\|^2 + \frac{\varepsilon^2}{E^2} \|I_h x^o\|_q^2$$

$$\leq (|A(I_h x^o - x^o)| + \varepsilon)^2 + \frac{\varepsilon^2}{E^2} c \|x^o\|_q^2$$

$$\leq (M\|I_h x^o - x^o\|_{-a} + \varepsilon)^2 + c\,\varepsilon^2 E^2$$

$$\leq (Mc_1 \varepsilon \|x^o\|_q + \varepsilon)^2 + c\,\varepsilon^2 E^2$$

$$\leq \bar{c}^2 \varepsilon^2 (1 + E^2)$$

where $\bar{c} \geq o$ is independent of ε and E. This implies

$$\|Ax^{\varepsilon,h} - y^{\varepsilon}\| \leq \bar{c}\,\varepsilon (1 + E^2)^{1/2} \;,\quad \|x^{\varepsilon,h}\|_q \leq \bar{c}\,(1 + E^2)^{1/2} \;,$$

$$\|Ax^{\varepsilon,h} - Ax^o\| \leq 2\bar{c}\,\varepsilon (1 + E^2)^{1/2} \;,$$

and
$$\|x^{\varepsilon,h} - x^o\|_o \leq \|x^{\varepsilon,h} - x^o\|_{-a}^{\frac{q}{q+a}} \|x^{\varepsilon,h} - x^o\|_q^{\frac{a}{q+a}}$$

$$\leq \|A(x^{\varepsilon,h} - x^o)\|_{-a}^{\frac{q}{q+a}} (2\bar{c}(1 + E^2)^{1/2})^{\frac{a}{q+a}}$$

$$\leq (2\bar{c}(1 + E^2)^{1/2})^{\frac{q}{q+a}} (2\bar{c}(1 + E^2)^{1/2})^{\frac{a}{q+a}} \varepsilon^{\frac{q}{q+a}}$$

$$= 2\bar{c}(1 + E^2)^{1/2} \varepsilon^{\frac{q}{q+a}} \;.$$

$\square$

Remarks 6.17

i) If we compare the estimates in Th.6.10 and in Th.6.16 we see that the parameter choice strategy $h = \varepsilon^{\frac{1}{q+a}}$ leeds to an "optimal" reconstruction result.

ii) As we see from the estimates in Th.6.10 and in Th.6.16, if the problem gets more ill-posed (a increases) $\frac{1}{q+a}$ increases. Thus, we come to the conclusion that the more ill-posed the problem is, the coarser the "mesh size" (parameter h) should be chosen.

iii) Examples for subspaces $(X_h)_{h>o}$ which satisfy the assumption (6.21) are given by finite element spaces.

Bibliographical comments

A unified treatment of the theory of generalized inverses is presented in GROETSCH [43] . A detailed discussion of Tikhonov's method is given in [47] . The content of Section 6.3 and 6.5 is an extract of NATTERER [76],[79] . Discrepancy principles are described by MOROZOV [74], ENGL, NEUBAUER [35] and GROETSCH [46] .

Chapter 7: Regularization by discretization

Here we consider the discretization of ill-posed problems by projection methods. We present error estimates and convergence results.

7.1 Discretization by projection methods.

Let X and Y be Hilbert spaces and let A be a bounded linear operator from X into Y. We consider methods for the approximate solution of the equation

$$(7.1) \qquad Ax = y.$$

A projection method to solve the equation (5.1) is defined as follows: Let $(X_h)_{h>o}$ and $(Y_h^*)_{h>o}$ be families of subspaces of X and Y respectively. Then the discrete problem is given as follows:

$$(7.2) \qquad \begin{array}{l} \text{Find } x_h \in X_h \text{ such that} \\[1ex] (Ax_h - y, w)_Y = 0 \text{ for all } w \in Y_h^* . \end{array}$$

We assume

$$(7.3) \qquad \dim X_h = \dim Y_h^* =: n_h \ , \quad n_h \in {\rm I\!N} .$$

To determine $x_h \in X_h$, we choose bases $\{u_1, \ldots, u_{n_h}\}$ and $\{v_1, \ldots, v_{n_h}\}$ in X_h and Y_h^* respectively, and represent x_h in the form

$$x_h = \sum_{j=1}^{n_h} c_j u_j .$$

Then (7.2) leads to the linear system

$$Mc = b$$

where

$$M = (m_{ij}), \ m_{ij} = (Au_j, v_i) \ , \ 1 \le i, j \le n_h \ ,$$

$$b = (b_i), \quad b_i = (y, v_i) \ , \quad 1 \le i \le n_h.$$

We assume that the equation $Mc = b$ is uniquely solvable, that is

$$(7.4) \qquad \det(Au_j, v_i) \ne 0.$$

Moreover, we assume

$$(7.5) \qquad A^{-1} \text{ exists }.$$

Remark 7.1

In the concrete situation where A is an integral operator the spaces X_h and Y_h^* are spaces of simple functions ((trigono-metric)polynomials, splines,...). Note that even if X_h admits a local basis as it is the case in ordinary finite element methods, in general the matrix M is not sparse due to the non-local nature of integral operators. *

In the following let the assumptions (7.3),(7.4),(7.5) are satisfied. Then the problems

$$\text{(G)} \quad \text{Given } y \in R(A), \text{ find } x \in X \text{ with } Ax = y$$

and

$$\text{(G}_h\text{)} \quad \text{Given } y \in Y, \text{ find } x_h \in X_h \text{ with } (Ax_h - y, w) = 0$$
$$\text{for all } w \in Y_h^*$$

are uniquely solvable and we can introduce linear bounded operators

$$P_h : X \longrightarrow X_h, \quad R_h : Y \longrightarrow X_h$$

by the definition

$$(AP_h u - Au, w) = 0 \qquad \text{for all } w \in Y_h^* ,$$
$$(7.6)$$
$$(AR_h z - z, w) = 0 \qquad \text{for all } w \in Y_h^* .$$

The denotation "projection method" becomes clear from the following

Lemma 7.2

We have $P_h = R_h A$, $P_h^2 = P_h$.

Proof: The first assertion follows immediately from the definition of P_h and R_h.

Let $x \in X$. We have

$$(AP_h x - Ax, w) = 0, \quad (AP_h^2 x - AP_h, w) = 0$$
$$\text{for all } w \in Y_h^* .$$

This implies

$$(AP_h^2 x - Ax, w) = 0 \quad \text{for all } w \in Y_h^*$$

which gives with assumption (7.4) $P_h^2 x = P_h x$.

$\square$

With these mappings P_h and R_h we can write the solution x_h of (G_h) as

$$x_h = R_h y \, , \quad x_h = P_h x, \quad \text{if} \quad Ax = y.$$

In the following we need the quantity

$$d(z, X_h) := \inf\{\|z - u\| \mid u \in X_h\} \, , \quad z \in X \, ,$$

which may be used to measure how well elements in X can be approximated by elements in X_h.

Now, consider again the usual setting:

$$x^o \in X \, , \quad y^o, y^\varepsilon \in Y \text{ with}$$

(7.7)

$$Ax^o = y^o, \quad \|y^o - y^\varepsilon\| \leq \varepsilon \quad (\varepsilon \geq o).$$

Then we may use the operator R_h as a reconstruction operator. We set

$$x^{\varepsilon, h} := R_h y^\varepsilon.$$

The main result concerning the reconstruction error $x^{\varepsilon, h} - x^o$ is given in

<u>Theorem 7.3</u>

We have

$$(7.8) \qquad \|x^{\varepsilon,h} - x^o\| \leq (1 + \|P_h\|)d(x^o, X_h) + \|R_h\|\varepsilon .$$

<u>Proof:</u> Since dim $X_h < \infty$ we know that there exists $z \in X_h$ with $\|z - x^o\| = d(x^o, X^h)$.

$$\|x^{\varepsilon,h} - x^o\| \leq \|P_h x^o - z\| + \|z - x^o\| + \|P_h x^o - x^{\varepsilon,h}\|$$

$$\leq \|P_h(x^o - z)\| + \|z - x^o\| + \|R_h A x^o - R_h y^\varepsilon\|$$

$$\leq \|P_h\|\|x^o - z\| + \|x^o - z\| + \|R_h\|\|y^o - y^\varepsilon\|$$

$$\leq (1 + \|P_h\|)d(x^o, X_h) + \|R_h\|\varepsilon. \qquad \square$$

In Section 7.3 we shall discuss specific projection methods. Here we look at the following example which is only of theoretical interest.

<u>Example 7.4</u>

Let A be a compact operator with singular system $(\sigma_j, e_j, f_j)_{j \in \mathbb{N}}$ and dense range $R(A)$. For $h > o$ let $n_h \in \mathbb{N}$ with

$$\sigma_{n_h} \geq h \quad , \quad \sigma_{n_h+1} \leq h$$

and put

$$X_h := \mathrm{span}\{e_1, \ldots, e_{n_h}\} \; , \; Y_h^* = \mathrm{span}\{f_1, \ldots, f_{n_h}\} \; .$$

Then if $\quad x = \sum_{j=1}^{\infty} (x, e_j)e_j, \quad y = \sum_{j=1}^{\infty} (y, f_j)f_j \quad$ we obtain

$$P_h x = \sum_{j=1}^{n_h} (x, e_j)e_j, \quad R_h y = \sum_{j=1}^{n_h} \sigma_j^{-1}(y, f_j)e_j.$$

Therefore

$$\|P_h\| = 1, \quad \|R_h\| = \sigma_{n_h}^{-1}, \quad h > o. \qquad *$$

7.2 Quasioptimality and robustness

Definition 7.5

1) The projection method (X_h, Y_h^*) is called <u>quasioptimal</u> if and only if

    i) $\lim_{h \to o} d(x, X_h) = 0$ for all $x \in X$;

    ii) $\exists c \geq o \quad \forall h > o \ (\|P_h\| \leq c)$.

2) The projection method (X_h, Y_h^*) is called <u>robust</u> if and only if there exists $c \geq o$ such that

$$\|R_h\| \leq c \, \sigma_h$$

where

$$\sigma_h := \sup \{\|u\| \ \|Au\|^{-1} \, \big| \, u \neq \Theta, \ u \in X_h\}.$$

The number σ_h describes the modul of continuity of $A\big|_{X_h}^{-1}$. Since $P_h\big|_{X_h} = I$ we obtain

$$\|R_h\| = \sup\{\|R_h y\| \ \|y\|^{-1} \, \big| \, y \in Y, \ y \neq \Theta\}$$

$$\geq \sup\{\|R_h Au\| \ \|Au\|^{-1} \, \big| \, u \in X_h, \ u \neq \Theta\}$$

$$= \sup\{\|u\| \ \|Au\|^{-1} \, \big| \, u \in X_h, \ u \neq \Theta\}$$

$$= \sigma_h \ .$$

Therefore robustness means that $(\|R_h\|)_{h>o}$ has the same asymptotics as $(\sigma_h)_{h>o}$. Notice that $\lim_{h \to o} \sigma_h = \infty$ if $\lim_{h \to o} d(x, X_h) = 0$ for all $x \in X$ and A^{-1} is unbounded.

In Ex. 7.4 a projection method is given which is both quasioptimal and robust. But this method is of no practical interest since a singular system has to be known.

Corollary 7.6

Suppose that the projection method (X_h, Y_h^*) is quasioptimal and robust. Then there exists a constant c, independent of h, such that

$$(7.9) \qquad \|x^{\varepsilon,h} - x^o\| \le c \, (d(x^o, X_h) + \sigma_h \varepsilon), \quad h > o \ .$$

Proof: This follows from Th.7.3 and Def.7.5. $\qquad\qquad\qquad\square$

The estimate (7.9) for the reconstruction error $x^{\varepsilon,h} - x^o$ is of the typical form if the problem is ill-posed: The term $d(x^o, X_h)$ is small if h is small (under appropriate assumptions on $(X_h)_{h>o}$) , the other term $\sigma_h \varepsilon$ is large if h is small since $\lim\limits_{h\to o} \sigma_h = \infty$ (see above). Thus, we like to choose the discretization parameter h in such a way that the bound for $\|x^{\varepsilon,h} - x^o\|$ becomes minimal. This is done explicitly in Section 7.4.

Now we want to look for criteria which imply the quasioptimality and robustness. First we prove a helpful lemma.

Lemma 7.7

Let U, V be Hilbert spaces and let $T \in B(U, V)$. Then

$$(7.10) \qquad \|T\| = \sup_{u \in U, u \notin N(T)} \ \inf_{w \in U, w \in N(T)^{\perp}} \frac{\|w\| \, \|Tu\|}{|(w, u)|}$$

Proof: Let b be the number on the right-hand side of (7.10). Let $u \in U, u \ne \theta$. Then if $w \in U$ we have

$$\frac{\|w\| \, \|Tu\|}{|(w, u)|} \ge \frac{\|w\| \, \|Tu\|}{\|w\| \, \|u\|} = \frac{\|Tu\|}{\|u\|}$$

which implies

$$\inf_{w \in U, w \in N(T)^{\perp}} \frac{\|w\| \, \|Tu\|}{|(w, u)|} \ge \frac{\|Tu\|}{\|u\|}$$

and $b \ge \|T\|$.

To prove the converse inequality $b \le \|T\|$ let $u \in U$, $u \notin N(T)$.
Then $u = u^1 + u^2$ with $u^1 \in N(T)$, $u^2 \in N(T)^\perp$, $u^2 \ne \Theta$. Hence

$$\inf_{w \in U, w \in N(T)^\perp} \frac{\|w\| \, \|Tu\|}{|(w,u)|} \;\le\; \frac{\|u^2\| \, \|Tu\|}{|(u^2,u)|} \;=\; \frac{|Tu^2|}{\|u^2\|} \;\le\; \|T\|.$$

This implies $b \le \|T\|$. $\qquad\qquad\qquad\square$

Theorem 7.8

We have

i) $\quad \|P_h\| = \sup\limits_{x \in X_h} \; \inf\limits_{v \in Y_h^*} \; \dfrac{\|A^*v\| \, \|x\|}{|(A^*v,x)|}$,

ii) $\quad \|R_h\| = \sup\limits_{x \in X_h} \; \inf\limits_{v \in Y_h^*} \; \dfrac{\|v\| \, \|x\|}{|(v,Ax)|}$.

<u>Proof:</u> We prove only ii) the proof of i) is similar.

By Lemma 7.7

$$\|R_h\| = \sup_{y \in Y,\, y \notin N(R_h)} \; \inf_{w \in Y,\, w \in N(R_h)^\perp} \frac{\|w\| \, \|R_h y\|}{|(w,y)|}$$

Since $N(R_h)^\perp = Y_h^*$ as one sees easily and since $R_h y = z$ if and
only if $(Az,v) = (y,v)$ for all $v \in Y_h^*$ (see (7.6)) we arrive at

$$\|R_h\| = \sup_{x \in R(R_h)} \; \inf_{v \in Y_h^*} \frac{\|v\| \, \|x\|}{|(v,Ax)|}$$

$$= \sup_{x \in R(R_h)} \; \inf_{v \in Y_h^*} \frac{\|v\| \, \|x\|}{|(v,Ax)|} \qquad\qquad \square$$

As we shall see in Section 7.3, in certain situations it is
easier to compute the norm of P_h and/or R_h in a norm different
of the given norm in X.

Lemma 7.9

Suppose that in X there are given additional norms $\|.\|_1$ and
$\|.\|_2$ such that with

$$\lambda_h := \sup\{\|z\|\ \|z\|_1^{-1} \mid z \in X_h,\ z \neq \Theta\}$$

the following properties hold:

i) $\exists c_1 \geq 0 \forall h > 0\ \forall z \in X_h\ (\|z\|_1 \leq c_1 \|z\|_2)$.

ii) $\exists c_2 \geq 0 \forall h > 0\ \forall z \in X \exists u \in X_h\ (\|z-u\| + \lambda_h \|z-u\|_2 \leq c_2 \|z\|)$.

iii) $\exists c_3 \geq 0 \forall h > 0\ \forall x \in X_h\ (\|P_h x\|_2 \leq c_3 \|x\|_2)$.

Then the projection method (X_h, Y_h^*) is quasioptimal.

Proof: Let $h > 0$, $z \in X$ and choose $u \in X_h$ according to
property ii).Then

$$\|z - P_h z\| \leq \|z - u\| + \|u - P_h z\|$$

$$\leq \|z - u\| + \frac{\|u - P_h z\|}{\|u - P_h z\|_1} \frac{\|u - P_h z\|_1}{\|u - P_h z\|_2} \|u - P_h z\|_2$$

$$\leq \|z - u\| + \lambda_h c_1 \|u - P_h z\|_2$$

$$\leq \|z - u\| + \lambda_h c_1 \|P_h (u - z)\|_2$$

$$\leq \|z - u\| + \lambda_h c_1 c_3 \|u - z\|_2$$

$$\leq \max(1, c_1 c_3)(\|z - u\| + \lambda_h \|u - z\|_2)$$

$$\leq \max(1, c_1 c_3) c_2 \|z\| = c_4 \|z\|$$

where $c_4 := \max(1, c_1 c_3) c_2$. This implies

$$\|P_h z\| \leq (1 + c_4)\|z\| \quad ,\ z \in X,$$

which shows $\|P_h\| \leq 1 + c_4$. $\qquad\qquad\qquad\qquad\qquad\square$

Remark 7.10

If A^{-1} is bounded then the projection method (X_h, Y_h^*) is
robust if and only if the method is quasioptimal. This follows

immediately from the identities

$$P_h = R_h A \ , \quad R_h = P_h A^{-1} \quad (h > o).$$

Therefore robustness plays a role only in ill-posed problems.

7.3 Specific methods

We want to consider three well-known projection methods. Let us begin with

The least squares method.

Choose $X_h \subset X$ and put $Y_h^* := AX_h$.

The associated problem (G_h) is equivalent with

$$\text{minimize } \tfrac{1}{2} \|Ax - y\|^2 \ .$$
$$x \in X_h$$

If we define norms $\|.\|_1$ and $\|.\|_2$ by

$$\|z\|_1 := \|z\|_2 := \|Az\| \ , \quad z \in X \ ,$$

then we obtain by using (7.6)

$$\frac{\|P_h x\|_2}{\|x\|_2} = \frac{(AP_h x, AP_h x)}{\|Ax\|^2} = \frac{\|Ax\|^2}{\|Ax\|^2} = 1 \ , \quad x \in X_h, \ x \neq \Theta \ ,$$

which shows that property iii) in Lemma 7.9 is satisfied (with $c_3 = 1$). Since condition i) in Lemma 7.9 holds, we conclude from this lemma that the following condition is sufficient for quasioptimality:

$$(7.11) \qquad \exists c \geq o \quad \forall h > o \quad \forall z \in X \ \exists u \in X_h$$

$$(\|z - u\| + \sigma_h \|A(z - u)\| \leq c \|z\|)$$

$$\text{where } \sigma_h = \sup \ \{\|v\|_X \mid \|Av\| \leq 1, \quad v \in X_h\}.$$

Since by Lemma 7.8

$$\|R_h\| = \sup_{x \in X_h} \ \inf_{z \in X_h} \ \frac{\|Az\| \ \|x\|}{(Az, Ax)}$$

$$\leq \sup_{x \in X_h} \frac{\|Ax\| \, \|x\|}{\|Ax\|^2} = \sigma_h$$

we have robustness without any additional assumption.

<u>The Ritz method:</u>

Here it is assumed $X = Y$, $A = A^*$.

Choose $X_h \subset X$ and put $Y_h^* := X_h$.

The associated problem (G_h) is equivalent to the problem

$$\min_{x \in X_h} \frac{1}{2} (x, Ax) - (y, x).$$

If we define norms $\|\cdot\|_1$ and $\|\cdot\|_2$ by

$$\|z\|_1 := \|z\|_2 := (z, Az)^{1/2}, \quad z \in X,$$

then we obtain from (7.6)

$$\sup_{x \in X} \frac{\|P_h x\|_2}{\|x\|_2} = 1$$

which shows together with Lemma 7.9 that the following condition is sufficient for quasioptimality:

$$(7.12) \quad \exists c \geq o \quad \forall h > o \quad \forall z \in X \quad \exists u \in X_h$$

$$(\|z - u\| + \lambda_h (z - u, A(z - u)))^{1/2} \leq c\|z\|)$$

$$\text{where } \lambda_h = \sup \{\|x\| \mid x \in X_h \, , \, (x, Ax)^{1/2} \leq 1\}.$$

Using Theorem 7.8 and the fact that A is selfadjoint we obtain

$$\|R_h\| = \sup_{x \in X_h} \inf_{v \in X_h} \frac{\|v\| \, \|x\|}{|(v, Ax)|}$$

$$= \sup_{x \in X_h} \inf_{v \in X_h} \frac{\|v\|}{\|A^* v\|} \cdot \frac{\|x\| \, \|A^* v\|}{|(v, Ax)|}$$

$$\leq \sup \left\{ \frac{\|v\|}{\|A^*v\|} \mid v \in X_h \right\} \cdot \sup_{x \in X_h} \inf_{v \in X_h} \frac{\|x\| \, \|A^*v\|}{|(v, Ax)|}$$

$$= \sigma_h \, \|P_h\|$$

which shows that the Ritz method is robust if it is quasioptimal. The condition (7.12) is therefore sufficient for robustness.

The generalized least squares method:

Let W be a Hilbert space and let $T \in B(Y,W)$.

Choose X_h and put $Y_h^* := T^*TAX_h$.

The associated problem (G_h) is equivalent to the problem

$$\underset{x \in X_h}{\text{minimize}} \; \frac{1}{2} \, \|T(Ax - y)\|^2 .$$

If we define norms $\|.\|_1$ and $\|.\|_2$ by

$$\|z\|_1 := \|Az\| \; , \; \|z\|_2 := \|TAz\| \; , \; z \in X \; ,$$

then we obtain by using (7.6)

$$\sup_{x \in X_h} \frac{\|P_h x\|_2}{\|x_2\|} = 1 \quad .$$

Therefore, by Lemma 7.9 the following conditions are sufficient for quasioptimality

$$(7.13) \quad \exists c_1 > 0 \quad \forall h > 0 \quad \forall z \in X_h \; (\|Az\| \leq c_1 \|TAz\|)$$

$$(7.14) \quad \exists c_2 \geq 0 \quad \forall h > 0 \quad \forall z \in X_h \quad \exists u \in X_h$$

$$(\|z - u\| + \sigma_h \, \|A(z - u)\| \leq c_2 \|z\|) \quad .$$

If (7.13) is satisfied the generalized least squares method is robust since

$$\|R_h\| = \sup_{x \in X_h} \inf_{v \in Y_h^*} \frac{\|v\| \, \|x\| \, \|Ax\|}{|(v, Ax)| \, \|Ax\|}$$

$$\leq \sup_{x \in X_h} \frac{\|x\|}{\|Ax\|} \quad \inf_{z \in X_h} \frac{\|T^*TAz\| \ \|Ax\|}{\|(T^*TAz, Ax)\|}$$

$$\leq \sigma_h \sup_{x \in X_h} \frac{\|T^*\| \ |TAx| \ |A\bar{x}|}{\|TAx\|^2}$$

$$\leq \sigma_h \ c_1^{-1} \ \|T^*\| \quad .$$

7.4 Asymptotic estimates

In this section we assume that the following a-priori information is given:

$$(7.15) \qquad x^o \in H_q, \quad \|x^o\|_q \leq E \quad (q \geq o, \ E \geq o);$$

here $(H_s)_{s \in \mathbb{R}}$ is a Hilbert scale constructed by an orthonormal basis $(e_j)_{j \in \mathbb{N}}$ in $H_o := X$ and a sequence $(\alpha_j)_{j \in \mathbb{N}}$ (see Section 5.1). In addition, we assume that the degree of ill-posedness is given by the following "smoothing" property:

There exists a > o such that

$$(7.16) \qquad m\|x\|_{-a} \leq \|Ax\| \leq M\|x\|_{-a} \ , \quad x \in H_o$$

with some constants $M \geq m > o$.

For the subspace X_h we take a space which satisfies the typical assumptions of finite element spaces:

$$\text{i)} \quad X_h \subset H_q \ , \quad \dim X_h < \infty.$$

$$(7.17) \quad \text{ii)} \quad \inf \{\|u - x\|_o + h^{-r}|u - x\|_{-r} \ | \ u \in X_h\} \leq h^s \ c_{r,s}\|x\|_s$$

$$\text{for all } x \in H_s \text{ where } -a \leq -r \leq s \leq q.$$

$$\text{iii)} \quad \|u\|_o \leq ch^{-a}\|u\|_{-a} \quad \text{for all } u \in X_h.$$

Here, $c_{r,s}, c$ are constants independent of h,u, and x. In the following, let the assumptions (7.3),(7.4),(7.5),(7.16) and (7.17) are satisfied. The main result in this chapter is (see Section 7.3)

<u>Theorem 7.11</u>

a) The least squares method is quasioptimal and robust. Moreover, the following estimate holds:

$$(7.18) \qquad \| x^{h,\varepsilon} - x^o \| \le c(h^q E + h^{-a}\varepsilon).$$

Here c is independent of h, E and ε.

b) The Ritz method is quasioptimal and robust provided (see (7.16))

$$(7.19) \qquad m\|x\|_{-\frac{a}{2}}^2 \le (x,Ax) \le M\|x\|_{-\frac{a}{2}}^2 \qquad \text{for all } x \in H_o.$$

If this is the case the estimate (7.18) holds.

c) The generalized least squares method is quasioptimal and robust provided the condition (7.13) is satisfied. If this is the case the estimate (7.18) holds.

<u>Proof:</u> Let us calculate σ_h: If $z \in X_h$, $z \ne \theta$, then

$$\| z \| \; \| Az \|^{-1} = \| z \|_o \| Az \|^{-1} \le \| z \|_o m^{-1} \| z \|_{-a}^{-1} \le m^{-1} \; ch^{-a} \| z \|_o \| z \|_o^{-1}$$

by assumption (7.16) and (7.17) iii). Therefore we have

$$\sigma_h = O(h^{-a}).$$

Using this fact and the assumptions (7.16),(7.17)ii) we obtain that the condition (7.11) is satisfied. Then we know that the least squares method is quasioptimal and robust. From assumption (7.17)ii) we obtain

$$d(x^o,X_h) = \inf \{\| u - x^o \|_o \, | u \in X_h\}$$

$$\le \inf \{\| u - x^o \|_o + h^{-a} \| u - x^o \|_{-a} | \, u \in X_h\}$$

$$\le c \, h^q \, \| x^o \|_q.$$

This implies the estimate (7.18) and part a) is proved.

In order to prove part b) it is sufficient to show that the condition (7.12) in section 7.3 is satisfied. But this condi-

tion is a simple consequence of (7.19) and (7.17)ii)
($r = \frac{a}{2}$, s = o). Part c) follows from the fact that (7.17)ii)
implies condition (7.14). $\square$

Balancing the terms in the estimate (7.18) we come to the
discretization parameter choice strategy

$$(7.20) \qquad h(\varepsilon) := (\frac{\varepsilon}{E})^{\frac{1}{q+a}}$$

and the bound

$$(7.21) \qquad \| x^{h(\varepsilon),\varepsilon} - x^{o} \| \leq 2\, c\, h^{\frac{q}{q+a}}\, E^{\frac{a}{q+a}}.$$

Comparing (7.21) with the estimate (6.10) we see that the
specific methods considered in Section 7.3 with an optimal
discretization parameter provide an optimal reconstruction.

<u>Remark 7.12</u>

The assumption (7.19) follows from assumption (7.16). The
proof of this result needs an interpolation theorem for operators in Hilbert scales.   *

Finally, let us consider the stability of the discretized
equation Mc = b which results from (7.2). The relevant parameter here is the condition number κ_h, the ratio of the
largest to the smallest eigenvalue of M.
We look only on the least squares method and assume that
the assumptions (7.3),(7.4),(7.5),(7.16) and (7.17) are satisfied. Let $u_1,\ldots,u_{n_h}$ be a basis of X_h. Then we have

$$M = (m_{ij}),\quad m_{ij} = (Au_j, Au_i),\quad 1 \leq i, j \leq n_h ,$$

$$b = (b_i)\ ,\quad b_i = (y^{\varepsilon}, Au_i)\ ,\quad 1 \leq i \leq n_h.$$

Then if $\lambda_{min} = \lambda_1 \leq \ldots \leq \lambda_{n_h} = \lambda_{max}$ are the eigenvalues of M
and if $x^{h,\varepsilon}$ has the representation $x^{h,\varepsilon} = \sum_{i=1}^{n_h} w_i^{h,\varepsilon} u_i$ we obtain

$$\lambda_{max} = \sup\{(w,Mw) \mid w \in \mathbb{R}^n, \|w\| \leq 1\}$$

$$\geq (w^{h,\varepsilon},Mw^{h,\varepsilon}) \; \|w^{h,\varepsilon}\|^{-2}$$

$$= \|Ax^{h,\varepsilon}\|^2 \; \|w^{h,\varepsilon}\|^{-2}$$

$$\geq m\|x^{h,\varepsilon}\|_{-a}^2 \; \|w^{h,\varepsilon}\|^{-2}$$

$$\geq m\,c\,h^{-2a} \|x^{\varepsilon,h}\|_o \; \|w^{h,\varepsilon}\|^{-2} \quad ,$$

$$\lambda_{min} = \inf \{(w,Mw) \mid w \in \mathbb{R}^N, \|w\| \leq 1\}$$

$$\leq (w^{h,\varepsilon},Mw^{h,\varepsilon}) \; \|w^{h,\varepsilon}\|^{-2}$$

$$= \|Ax^{h,\varepsilon}\|^2 \; \|w^{h,\varepsilon}\|^{-2}$$

$$\leq M\|x^{h,\varepsilon}\|_{-a}^2 \; \|w^{h,\varepsilon}\|^{-2}$$

$$\leq M\overline{c} \; \|x^{h,\varepsilon}\|_o^2 \; \|w^{h,\varepsilon}\|^{-2}$$

This gives a lower bound for the condition number:

$$(7.22) \quad \kappa_h \geq c'h^{-2a} \quad ;$$

here c' is a constant independent of h.

Bibliographical comments

The numerical solution of ill-posed problems by projection methods is treated in NATTERER [77],[78] and RICHTER [87]. The notion of robustness has been introduced in [77]. HSIAO, WENDLAND [57],[103] proved quasioptimality of Galerkin-collocation methods for certain integral operators, moment methods have been considered by NASHED [75]. More on condition numbers of matrices arising from discretization methods can be found in WING [106].

PART III

LEAST SQUARES SOLUTIONS OF SYSTEMS OF
LINEAR EQUATIONS

In this part we discuss problems of numerical linear algebra which are of interest in the solution of inverse problems: least squares problems, inconsistent systems of linear equations, decomposition algorithms.

Chapter 8: Least squares problems

In this chapter we are concerned with the least squares solution of systems of linear equations, i.e. the minimization of $\|Ax - y\|$ where $A \in \mathbb{R}^{m,n}$, $y \in \mathbb{R}^m$ and $\| \ \|$ denotes the euclidean distance. Our aim is to establish the mathematical material to discuss the computation of least squares solutions in the next chapter. The results are developed independently of results in Part II.

8.1 The singular value decomposition of a matrix

Let $M = (m_{ij}) \in \mathbb{R}^{m,n}$, $u, v \in \mathbb{R}^k$, $\alpha_1, \ldots, \alpha_1 \in \mathbb{R}$, $u_1, \ldots,$ $u_1 \in \mathbb{R}^k$. We denote by

I	the identity matrix,
M^t	the transpose of M,
$\|u\|$	the euclidean norm of u,
$u^t v$	the euclidean inner product of u and v,
$\|M\|$	the norm max $\{\|Mx\| \mid x \in \mathbb{R}^n, \|x\| \leq 1\}$,
$\|M\|_F$	the Frobenius norm of M, that is

$$\| M \|_F := \left(\sum_{i,j} |m_{ij}|^2 \right)^{1/2},$$

$\langle u_1,\ldots,u_1 \rangle$ the linear span of $u_1,\ldots,u_1$,

$\mathrm{diag}(\alpha_1,\ldots,\alpha_1)$ the diagonal matrix in $\mathbb{R}^{p,1}$ with elements $\alpha_1,\ldots,\alpha_1$ in its diagonal $(p \geq 1)$,

$(u_1|\ldots|u_1)$ the matrix in $\mathbb{R}^{k,1}$ with columns $u_1,\ldots,u_1$.

The main result in this section is

Theorem 8.1

Let $A \in \mathbb{R}^{m,n}$. Then there exist matrices $U \in \mathbb{R}^{m,m}$ and $V \in \mathbb{R}^{n,n}$ and real numbers $\sigma_1 \geq \ldots \geq \sigma_n \geq 0$ such that

$$(8.1) \quad A = UDV^t, \quad U^tU = UU^t = I, \quad V^tV = VV^t = I,$$

where $D = \mathrm{diag}(\sigma_1,\ldots,\sigma_n) \in \mathbb{R}^{m,n}$.

__Proof:__ Since the matrix $A^tA \in \mathbb{R}^{n,n}$ is symmetric and positive semidefinite there exist real numbers $\sigma_1 \geq \ldots \geq \sigma_n \geq 0$ such that $\sigma_1^2,\ldots,\sigma_n^2$ are the eigenvalues of A^tA.

Let $v_1,\ldots,v_n \in \mathbb{R}^n$ be an orthonormal family of eigenvectors of A^tA associated to $\sigma_1^2,\ldots,\sigma_n^2$ and let $r \leq n$ with $\sigma_r > 0$, $0 = \sigma_{r+1} = \ldots = \sigma_n = 0$. Then if we set $V_1 := (v_1|\ldots|v_r)$, $V_2 := (v_{r+1}|\ldots|v_n)$, $V := (v_1|\ldots|v_n)$ and $D_1 := \mathrm{diag}(\sigma_1|\ldots|\sigma_r) \in \mathbb{R}^{r,r}$ we have

$$D_1^{-1} = \mathrm{diag}(\sigma_1^{-1},\ldots,\sigma_r^{-1}), \quad V_1^tA^tAV_1 = D_1^2,$$

$$D_1^{-1}V_1^tA^tAV_1D_1^{-1} = I \in \mathbb{R}^{r,r}, \quad V_2^tA^tAV_2 = \theta,$$

$$AV_2 = \theta .$$

Let $U_1 := A_1V_1D_1^{-1} \in \mathbb{R}^{m,r}$. Then if we choose a matrix

$U_2 \in \mathbb{R}^{m,m-r}$ such that $U := (U_1 | U_2) \in \mathbb{R}^{m,m}$ and $U^t U = U U^t = I$ we obtain

$$U^t A V = \begin{pmatrix} U_1^t A V_1 & U_1^t A V_2 \\ U_2^t A V_1 & U_2^t A V_2 \end{pmatrix} = \begin{pmatrix} D_1 & \Theta \\ \Theta & \Theta \end{pmatrix}.$$

$\square$

Remarks 8.2

1) The number r in the proof of Theorem 8.1 is the rank of A.

2) As we know, the eigenvalues of $A^t A$ are uniquely determined. Therefore the numbers $\sigma_1, \ldots, \sigma_n$ are uniquely determined too.

3) Of course, Th. 8.1 follows immediately from Th. 4.14 but presents the results in this chapter selfcontained.

$*$

By 2) in Remark 8.2 the following definition makes sense.

Definition 8.3

Let $A \in \mathbb{R}^{m,n}$ with $m \geq n$. A decomposition of the form (8.1) is called a <u>singular value decomposition of A</u> and the numbers $\sigma_1, \ldots, \sigma_n$ are called the <u>singular values of A</u>.

Corollary 8.4

Suppose that $A \in \mathbb{R}^{m,n}$ and let $A = UDV^t$ be a singular value decomposition of A where $U = (u_1 | \ldots | u_m)$, $V = (v_1 | \ldots | v_n)$ and $D = \mathrm{diag}(\sigma_1, \ldots, \sigma_r, 0, \ldots, 0)$, $\sigma_1 \geq \ldots \geq \sigma_r > 0$. Then

i) $A^t A \, v_i = \sigma_i^2 \, v_i$, $A A^t \, u_i = \sigma_i^2 \, u_i$, $1 \leq i \leq r$;

ii) $A \, v_i = \sigma_i \, u_i$, $A^t \, u_i = \sigma_i \, v_i$, $1 \leq i \leq r$;

iii) $\operatorname{rank}(A) = r$;

iv) $< v_1, \ldots, v_r > = N(A)^\perp$, $< v_{r+1}, \ldots, v_r > = N(A)$;

v) $< u_1, \ldots, u_r > = R(A)$, $< u_{r+1}, \ldots, u_m > = R(A)^\perp$;

vi) $\| A \| = \sigma_1 = \| A^t \|$; $\| A^+ \| = \dfrac{1}{\sigma_r}$.

Proof: All assertions are immediate consequences of the singular value decomposition. Therefore the proof may be left to the reader.

$\square$

Corollary 8.5

Suppose that $A \in \mathbb{R}^{m,n}$ has rank r. Then there exist matrices $B \in \mathbb{R}^{m,r}$ and $C \in \mathbb{R}^{r,n}$ and real numbers $\sigma_1 \geq \ldots \geq \sigma_r > 0$ such that

$$A = BC, \quad B^t B = I, \quad CC^t = \operatorname{diag}(\sigma_1^2, \ldots, \sigma_r^2),$$

(8.2)

$$\operatorname{rank}(B) = \operatorname{rank}(C) = r.$$

Proof: Without loss of generality assume $m \geq n$. Let $A = UDV^t$ be a singular value decomposition of A:

$$D = \operatorname{diag}(\sigma_1, \ldots, \sigma_r, 0, \ldots, 0), \quad U = (u_1 | \ldots | u_m),$$

$$V = (v_1 | \ldots | v_n).$$

Then if we set $B := (u_1 | \ldots | u_r)$, $C_1 := (v_1 | \ldots | v_r)$,
$C := \operatorname{diag}(\sigma_1, \ldots, \sigma_r) C_1$ we have the desired decomposition. $\square$

The following result is concerned with the perturbation of eigenvalues of a symmetric matrix. We need this result to understand how the singular values of a matrix can be affected by (small) perturbations.
Let us denote by

$$V_j^n := \{ V \subset \mathbb{R}^n \mid V \text{ is a subspace with } \dim V \leq j \}.$$

Theorem 8.6

Suppose that $C \in \mathbb{R}^{n,n}$ is a symmetric matrix with eigenvalues $\lambda_1 \geq \ldots \geq \lambda_n$. Then for $k = 1, \ldots, n$

$$\lambda_k = \min_{V \in V^n_{n-k+1}} \; \max_{x \in V, \|x\|=1} \; x^t C x = \max_{V \in V^n_k} \; \min_{x \in V, \|x\|=1} \; x^t C x \;.$$

Proof: Let $v_1, \ldots, v_n$ be an orthonormal system of eigenvectors associated with $\lambda_1, \ldots, \lambda_n$.

Let $V \in V^n_{n-k+1}$. Then $\dim V \cap \langle v_1, \ldots, v_k \rangle \geq 1$ and if

$$x = \sum_{i=1}^{k} \alpha_i v_i \in V \cap \langle v_1, \ldots, v_k \rangle \text{ with } \|x\| = 1 \text{ we obtain}$$

$$x^t C x = \sum_{i=1}^{k} \alpha_i^2 \lambda_i \geq \lambda_k \;.$$

This shows $\lambda_k \leq \min\limits_{V \in V^n_{n-k+1}} \; \max\limits_{x \in V, \|x\|=1} \; x^t C x.$

On the other hand, we have for each $x = \sum\limits_{i=k}^{n} \alpha_i v_i$ in $V := \langle v_k, \ldots, v_n \rangle$ with $\|x\| = 1$

$$x^t C x = \sum_{i=k}^{n} \alpha_i^2 \lambda_i \leq \lambda_k,$$

that is

$$\lambda_k \geq \min_{V \in V^n_{n-k+1}} \; \max_{x \in V, \|x\|=1} \; x^t C x \;.$$

Thus the first identity is proved. If we apply this identity to $-C$ we obtain the second identity.

$\square$

Theorem 8.7

Suppose that $C, F \in \mathbb{R}^{n,n}$ are symmetric matrices and let $\lambda_1 \geq \ldots \geq \lambda_n$, $\gamma_1 \geq \ldots \geq \gamma_n$ and $\mu_1 \geq \ldots \geq \mu_n$ be the eigenvalues of C,F and C + F respectively. Then we have for all $k \in \{1,\ldots,n\}$:

i) $\lambda_k + \gamma_n \leq \mu_k \leq \lambda_k + \gamma_1$;

ii) $|\lambda_k - \mu_k| \leq \|F\|$.

__Proof:__ Let $v_1,\ldots,v_n$ be an orthonormal system of eigenvectors of C.

i) We have for every $V \in V^n_{n-k+1}$

$$\mu_k \leq \max_{x \in V, \|x\| = 1} x^t(C+F)x$$

$$\leq \max_{x \in V, \|x\| = 1} x^t C x + \max_{x \in V, \|x\| = 1} x^t F x.$$

If we apply this to the space $V := \langle v_k,\ldots,v_n\rangle$ we obtain

$$\mu_k \leq \lambda_k + \max_{x \in V, \|x\| = 1} x^t F x = \lambda_k + \gamma_1 \ .$$

In the same manner we obtain the other inequality.

ii) From i) we have $\mu_k - \lambda_k \leq |\gamma_1|, \ \lambda_k - \mu_k \leq -\gamma_n \leq |\gamma_n|$. Since $|\gamma_1|,|\gamma_n| \leq \|F\|$ the desired result follows.

$\square$

__Theorem 8.8__

Let $A \in \mathbb{R}^{m,n}$ and let $\sigma_1 \geq \ldots \geq \sigma_n$ be the singular values of A. Then for all $k \in \{1,\ldots,n\}$

$$\sigma_k = \min_{V \in V^n_{n-k+1}} \ \max_{x \in V, \|x\| = 1} \|Ax\| = \max_{V \in V^n_k} \ \min_{x \in V, \|x\| = 1} \|Ax\| \ .$$

__Proof:__ Apply Theorem 8.6 to $C := A^t A$.

$\square$

In the sequel it is convenient to denote the singular values of a matrix $C \in \mathbb{R}^{m,n}$ by

$$\sigma_1(C), \ldots, \sigma_n(C)$$

where we always assume that these numbers satisfy

$$\sigma_1(C) \geq \ldots \geq \sigma_n(C) \ .$$

Theorem 8.9

Let $A, F \in \mathbb{R}^{m,n}$. Then for all $k \in \{1, \ldots, n\}$

$$|\sigma_k(A) - \sigma_k(A+F)| \leq \|F\| \ .$$

Proof: Follows immediately from Theorem 8.8 in the same manner as the result in Theorem 8.7 from Theorem 8.6.

$\square$

Illustration 8.10

Let $A := \begin{pmatrix} 1 & -1 \\ 0 & 0 \end{pmatrix}$, $F := \begin{pmatrix} 0 & 0 \\ 0 & 0.1 \end{pmatrix}$.

A simple calculation shows [*)]

$$\|F\| = 0.1$$

$$\sigma_1(A) \doteq 1.4142 \qquad \sigma_1(A+F) \doteq 1.4160$$

$$\sigma_2(A) \doteq 0.0000 \qquad \sigma_2(A+F) \doteq 0.0707$$

8.2 The pseudo-inverse

We consider the problem of finding solutions to the equation

$$(8.3) \qquad Ax = y$$

[*)] By the symbol $a \doteq t_1 \cdot s_1 \ldots s_n$ we mean $|a - t_1 \cdot s_1 \ldots s_n| \leq \frac{1}{2} 10^{-n}$.

where $A \in \mathbb{R}^{m,n}$ and $y \in \mathbb{R}^m$ are given. If $m = n$ and A is nonsingular, there is a unique solution, given by $x = A^{-1}y$. In the general case, when A may be singular or rectangular $(m \neq n)$, there may be none, one or a infinite number of solutions. One would like to be able to find a matrix $G \in \mathbb{R}^{n,m}$ such that $x := Gy$ is a solution of (8.3) in a generalized sense. Such a matrix should have the property that if $y \in R(A)$, i.e. $y = Az$, $z \in \mathbb{R}^n$, we have

$$AGy = AGAz = y \ .$$

This is satisfied if the matrix G has the following property

$$(8.4) \qquad AGA = A \ .$$

It is the aim of this section to construct such a matrix.

Let $T_A : \mathbb{R}^n \to \mathbb{R}^m$ be the linear map defined by the matrix $A \in \mathbb{R}^{m,n}$, i.e.

$$T_A x := Ax \ , \ x \in \mathbb{R}^n \ .$$

Then the linear mapping

$$T_A\Big|_{N(A)^\perp}^{-1} \ : \ R(A) \longrightarrow N(A)^\perp$$

is welldefined and the matrix $G \in \mathbb{R}^{n,m}$ which represents this linear mapping satisfies the identity (8.4). This suggests the following

<u>Definition 8.11</u>

Let $A \in \mathbb{R}^{m,n}$. Define the linear map $T_A^+ : \mathbb{R}^m \to \mathbb{R}^n$ by

$$T_A^+ y := \begin{cases} \Theta & , \text{ if } y \in R(A)^\perp \\ T_A\Big|_{N(A)^\perp}^{-1} y & , \text{ if } y \in R(A) \ . \end{cases}$$

Then the matrix $A^+ \in \mathbb{R}^{n,m}$ which represents this linear map T_A^+ is called the <u>pseudo-inverse of A</u> (i.e. $T_A^+ = T_{A^+}$).

It is clear that the pseudo-inverse A^+ of A is the ordinary inverse A^{-1} if $A \in \mathbb{R}^{n,n}$ and A is nonsingular. But the reader should be warned about certain facts that are true in the case of ordinary inverses while they are not true for pseudo-inverses.

Example 8.12

Let $A := \begin{pmatrix} 1 & -1 \\ 0 & 0 \end{pmatrix}$. Then $R(A^t)$ is spanned by the vectors $\begin{pmatrix} 1 \\ -1 \end{pmatrix}$ and $\begin{pmatrix} 0 \\ 0 \end{pmatrix}$, and a basis of $N(A)^\perp = R(A^t)$ is given by the vector $\begin{pmatrix} 1 \\ -1 \end{pmatrix}$. Now $A \begin{pmatrix} 1 \\ -1 \end{pmatrix} = \begin{pmatrix} 2 \\ 0 \end{pmatrix}$ and therefore $A^+\begin{pmatrix} 2 \\ 0 \end{pmatrix} = \begin{pmatrix} 1 \\ -1 \end{pmatrix}$. Using the identity $R(A)^\perp = N(A^t)$ we see that a basis of $R(A)^\perp$ is given by the vector $\begin{pmatrix} 0 \\ 1 \end{pmatrix}$; therefore $A^+\begin{pmatrix} 0 \\ 1 \end{pmatrix} = \begin{pmatrix} 0 \\ 0 \end{pmatrix}$.
This gives us

$$A^+ \begin{pmatrix} 2 & 0 \\ 0 & 1 \end{pmatrix} = \begin{pmatrix} 1 & 0 \\ -1 & 0 \end{pmatrix}, \quad A^+ = \begin{pmatrix} 1 & 0 \\ -1 & 0 \end{pmatrix} \begin{pmatrix} 2 & 0 \\ 0 & 1 \end{pmatrix}^{-1}$$

$$= \frac{1}{2} \begin{pmatrix} 1 & 0 \\ -1 & 0 \end{pmatrix}.$$

Thus, we have

$$A^2 = A, \quad (A^+)^2 \neq A^+. \qquad\qquad *$$

From the example above we may learn two things: First, in general it is not true that $(AB)^+ = B^+A^+$ even if $AB = BA$, and second, how to calculate A^+ from A (for small m,n).

Before we show how the singular value decomposition can be used for the computation of the pseudo-inverse we give an equivalent definition of the pseudo-inverse. To formulate this we need the following notation:
If S is a closed subspace of a Hilbert space $(\mathbb{R}^n)$ we denote by P_S the orthogonal projector onto S. By the projection theorem:

$$z = P_S x \quad \text{iff} \quad (z-x,u) = 0 \text{ for all } u \in S;$$

here $(.,.)$ is the inner product in the Hilbert space $(\mathbb{R}^n)$.

<u>Theorem 8.13</u>

Suppose that $A \in \mathbb{R}^{m,n}$ and $G \in \mathbb{R}^{n,m}$. Then the following conditions are equivalent:

a) $G = A^+$.

b) $AG = P_{R(A)}$, $GA = P_{R(G)}$.

c) $AGA = A$, $GAG = G$, $(AG)^t = AG$, $(GA)^t = GA$.

<u>Proof</u>:

a) $\Rightarrow$ b) From the definition of A^+ we obtain the identities

$$AA^+y = \theta \quad \text{if } y \in R(A)^\perp, \quad AA^+y = y \quad \text{if } y \in R(A),$$
$$A^+Ax = \theta \quad \text{if } x \in R(A^+)^\perp, \quad A^+Ax = x \quad \text{if } x \in R(A^+),$$

which imply the properties in b).

b) $\Rightarrow$ c) Since projectors are symmetric it is immediately clear that the conditions

$$(AG)^t = AG, \quad (GA)^t = GA$$

are satisfied.
The identities

$$AGA = A, \quad GAG = G$$

follow from

$$AGA = P_{R(A)}A, \quad GAG = P_{R(G)}G.$$

c) $\Rightarrow$ b) We have

$$(AG)(AG) = AG, \quad (AG)^t = AG,$$

and this shows that AG is an orthogonal projector. From

$$R(A) = R(AGA) \subset R(AG) \subset R(A)$$

we obtain

$$R(A) = R(AG), \quad AG = P_{R(A)}.$$

In the same manner we can prove $GA = P_{R(G)}$.

b) $\Rightarrow$ a) Since we know that the pseudo-inverse A^+ exists and satisfies the conditions in b) and c) we have only to show that a solution of the matrix equations in c) is unique. Suppose that $G := G_1$ and $G := G_2$ satisfy the equations in c). If $V := G_1 - G_2$ we have

$$AVA = \Theta \ , \ (AV)^t = AV \ , \ (VA)^t = VA.$$

This implies

$$(AV)^t AV = (AVA)V = \Theta \ , \ (VA)^t VA = V(AVA) = \Theta$$

which shows

$$\Theta = AV = AG_1 - AG_2 \ , \ \Theta = VA = G_1A - G_2A.$$

Thus

$$G_1 = G_1AG_1 = G_2AG_1 = G_2AG_2 = G_2 \ .$$

$\Box$

Example 8.14

Let $A \in \mathbb{R}^{m,n}$. Then we have

$$A^+ = \begin{cases} (A^tA)^{-1} A^t & \text{, if rank } (A) = n \\ A^t(AA^t)^{-1} & \text{, if rank } (A) = m \ . \end{cases}$$

This can be shown by verifying the conditions c) in Theorem 8.13.

$*$

Theorem 8.15

Let $A \in \mathbb{R}^{m,n}$ and let $A = UDV^t$ be a singular value decomposition of A. Then

$$D^+ = \text{diag} (\sigma_1^+, \ldots, \sigma_n^+) \ , \ A^+ = VD^+U^t$$

where $\sigma_1 \geq \ldots \geq \sigma_n$ are the singular values of A and

$$\sigma_i^+ = \begin{cases} \sigma_i^{-1} & \text{, if } \sigma_i > 0 \\ 0 & \text{, if } \sigma_i = 0 \end{cases} \ , \ 1 \leq i \leq n \ .$$

Proof: We have

$$D = \text{diag} (\sigma_1, \ldots, \sigma_n)$$

and

$$D^+ = \text{diag} (\sigma_1^+, \ldots, \sigma_n^+)$$

by Ex. 8.14. The identity $A^+ = VD^+U^t$ is proved by verifying the identities c) in Th. 8.13 for $G := VD^+U^t$. $\quad\Box$

Now we list some of the basic properties of pseudo-inverses. With Th. 8.13 it is easy to verify the assertions and thus, the proofs are left as exercises to the reader.

<u>Corollary 8.16</u>

Suppose that $A \in \mathbb{R}^{m,n}$. Then

i) $R(A) = R(AA^+) = R(AA^t)$

$\quad R(A^+) = R(A^t) = R(A^+A) = R(A^tA)$;

ii) $R(A)^\perp = N(A^+) = N(A^t) = N(AA^+)$;

iii) $(A^+)^+ = A,\quad (A^+)^t = (A^t)^+$;

iv) $A^t = A^tAA^+ = A^+AA^t,\quad (A^tA)^+ = A^+(A^t)^+$;

v) If $\lambda \in \mathbb{R}$ then $(\lambda A)^+ = \lambda^+A^+$ where

$\quad \lambda^+ = \lambda^{-1}$ if $\lambda \neq 0$ and $= 0$ if $\lambda = 0$.

8.3 Least squares solutions

We consider again the equation

(8.5) $$Ax = y$$

where $A \in \mathbb{R}^{m,n}$ and $y \in \mathbb{R}^m$. Using the pseudo-inverse A^+ of A we obtain a generalized solution A^+y of (8.5). In this section we shall see that the <u>pseudo-solution</u> A^+y is also a least squares solution, i.e. a vector that makes $Ax - y$ in the euclidean norm as small as possible.

<u>Theorem 8.17</u>

Let $A \in \mathbb{R}^{m,n}$ and $y \in \mathbb{R}^m$. Then

a) The problem

(8.6) $$\min \{ \|Ax - y\| \mid x \in \mathbb{R}^n \}$$

has always a solution $\bar{x}$.

b) The solution of (8.6) is uniquely determined if and only if $N(A) = \{\theta\}$.

c) Every solution $\bar{x}$ of (8.6) is a solution of the equation

(8.7)
$$A^t Ax = A^t y.$$

<u>Proof</u>: Let y_1 and y_2 be the projections of y in $R(A)$ and $R(A)^\perp$ respectively, i.e. $y = y_1 + y_2$, $y_1 \in R(A)$, $y_2 \in R(A)^\perp$, $y_1^t y_2 = 0$.

a) Since $y_1 - Ax \in R(A)$ for every $x \in \mathbb{R}^n$ we have by the Pythagorean law

$$\|Ax - y\|^2 = \|Ax - y_1\|^2 + \|y_2\|^2 \quad , \; x \in \mathbb{R}^n.$$

Thus $\bar{x}$ will be a solution of (8.6) if and only if $\bar{x}$ is a solution of $Ax = y_1$. Since $y_1 \in R(A)$ there exists a solution $\bar{x}$ of $Ax = y_1$.

b) This assertion follows from the fact that the solution of $Ax = y_1$ is uniquely determined if and only if $N(A) = \{\theta\}$.

c) We have $A\bar{x} = y_1 \in R(A)$, $y - A\bar{x} \in R(A)^\perp$. Since $R(A) = N(A^t)^\perp$ we obtain $\theta = A^t y_2 = A^t(y - A\bar{x})$.
$\qquad\qquad\qquad\qquad\qquad\qquad\qquad\qquad\qquad\qquad\qquad\qquad$ $\square$

<u>Definition 8.18</u>

Let $A \in \mathbb{R}^{m,n}$ and $y \in \mathbb{R}^m$. A vector $\bar{x} \in \mathbb{R}^n$ is called a <u>least squares solution</u> of the equation $Ax = y$ if $\bar{x}$ is a solution of the normal equation $A^t Ax = A^t y$.

<u>Remarks 8.19</u>

1) If $y \in R(A)$, then the notions of a solution and a least squares solution obviously coincide.

2)   Instead of problem (8.6) we can consider

$$\min \ \{ \ \|Ax - y\|_* \ | \ x \in \mathbb{R}^n \}$$

where $\|\cdot\|_*$ is any norm in $\mathbb{R}^n$. Different norms render different optimum solutions and therefore different generalized solutions of the equation $Ax = y$. Minimization in the l_1-norm or in the l_∞-norm is complicated by the fact that the function $x \longmapsto \|Ax - y\|_*$ is not differentiable for those norms.   *

From Theorem 8.17 it is clear that the set

$$L(y) \ := \ \{\bar{x} \in \mathbb{R}^n \ | \ \bar{x} \ \text{is a least squares solution of } Ax = y\}$$

is a nonempty closed convex subset of $\mathbb{R}^n$. Therefore, by the projection theorem, there exists a   uniquely determined vector $x^* \in L(y)$ such that

$$\|x^*\| \ = \ \min \ \{ \|\bar{x}\| \ | \ \bar{x} \in L(y) \}.$$

Definition 8.20

Let $A \in \mathbb{R}^{m,n}$ and $y \in \mathbb{R}^m$. Then a least squares solution of $Ax = y$ is called a <u>minimal norm solution</u> to $Ax = y$ if and only if

$$\|x\| \ = \ \min\{\|\bar{x}\| \ |\bar{x} \ \text{least squares solution of } Ax = y\}.$$

The next result shows that the minimal norm solution can be computed using the pseudo-inverse.

Theorem 8.21

Let $A \in \mathbb{R}^{m,n}$ and $y \in \mathbb{R}^m$. Then $x := A^+y$ is the uniquely determined minimal norm solution of $Ax = y$.

<u>Proof:</u> For every $x \in \mathbb{R}^n$ we have

$$\| Ax - y \|^2 = \| (Ax - AA^+y) + (AA^+ - I)y \|^2$$

$$= \| Ax - AA^+y \|^2 + \| (AA^+ - I)y \|^2$$

$$\geq \| Ax - AA^+y \|^2$$

since $AA^+ - I$ is the orthogonal projector onto $R(A)^\perp$ by Th. 8.13. This shows that $x := A^+y$ is a least squares solution and that every least squares solution of $Ax = y$ is a solution of $Ax = AA^+y$. But solutions of $Ax = AA^+y$ are of the form

$$x = P_{R(A)}x + (I - P_{R(A)})x = A^+Ax + (x - A^+Ax)$$

$$= A^+AA^+y + (x - A^+y) = A^+y + (x - A^+y)$$

which implies

$$\| x \|^2 = \| A^+y \|^2 + \| x - A^+y \|^2 \geq \| A^+y \|^2 .$$

From this we conclude that $x = A^+y$ is the uniquely determined minimal norm solution. $\square$

In the sequel we shall call a minimal norm solution also a <u>pseudo-solution</u>. This is suggested by the result of Th. 8.21.

8.4 Perturbation results

Now we come to the important question how errors in the data A and y of the equation $Ax = y$ propagate on the pseudo-solution $x = A^+y$; in other words, we compare the pseudo-solution of

$$(8.8) \qquad\qquad (A + F)x = y + u$$

with the pseudo-solution of

$$(8.9) \qquad\qquad Ax = y .$$

Thus, we have to study the question how the operation

$$^+ : \mathbb{R}^{m,n} \longrightarrow \mathbb{R}^{n,m} , \; M \longrightarrow M^+$$

operates.

<u>Lemma 8.22</u>

Suppose that $M \in \mathbb{R}^{m,m}$. If $\|M\| < 1$ then $(I - M)^{-1}$ exists and we have

i) $(I - M)^{-1} = \sum\limits_{k=0}^{\infty} M^k$;

ii) $\|(I - M)^{-1}\| \leq \dfrac{1}{1-\|M\|}$, $\|I - (I - M)^{-1}\| \leq \dfrac{\|M\|}{1-\|M\|}$.

<u>Proof:</u> The assertions follow immediately from the following identities which hold for all $s \in \mathbb{N}$:

$$(I - M) \left(\sum\limits_{k=0}^{s} M^k \right) = \left(\sum\limits_{k=0}^{s} M^k \right) (I - M) = (I - M^{s+1}) .$$

$\square$

<u>Theorem 8.23</u>

Let $A, F \in \mathbb{R}^{m,m}$ and assume that A is nonsingular and $\|A^{-1}\| \cdot \|F\| < 1$. Then $(A + F)^{-1}$ exists and if

$$Ax = y, \quad (A + F)\tilde{x} = y + u$$

for vectors $x, \tilde{x}, y, u \in \mathbb{R}^m, y \neq 0$, we have

$$(8.10) \qquad \frac{\|x-\tilde{x}\|}{\|x\|} \leq \frac{\|A^{-1}\| \cdot \|A\|}{1-\|A^{-1}\| \cdot \|A\|} \left(\frac{\|F\|}{\|A\|} + \frac{\|u\|}{\|y\|} \right) .$$

<u>Proof:</u> $(A + F)^{-1} = (I + A^{-1}F)^{-1}A^{-1}$ exists and we have

$$\| (I + A^{-1}F)^{-1}\| \leq \frac{1}{1-\|A^{-1}\| \cdot \|F\|}$$

$$\| I - (I + A^{-1}F)^{-1}\| \leq \frac{\|A^{-1}\| \cdot \|F\|}{1-\|A^{-1}\| \cdot \|F\|}$$

by Lemma 8.22. Now, the estimate (8.10) follows immediately from

143

$$x - \tilde{x} = x - (A + F)^{-1}y - (A + F)^{-1}u$$

$$= (I - (A + F)^{-1}A)x - (A + F)^{-1}u$$

$$= (I - (I + A^{-1}F)^{-1})x - (I + A^{-1}F)^{-1}A^{-1}u$$

and

$$\|y\| = \|Ax\| \leq \|A\| \, \|x\| . \qquad \square$$

The number

$$\kappa(A) := \|A\| \cdot \|A^{-1}\|$$

is called the <u>condition number of A.</u> $\kappa(A)$ measures the sensitivity of the solution x of $Ax = y$ to changes in the data A and y.

Now, we come back to the problem when A is a rectangular matrix. Our first result shows how very different from A^+ the pseudo-inverse of the perturbed matrix $A + F$ can be.

<u>Lemma 8.24</u>

Let $A, F \in \mathbb{R}^{m,n}$ and assume rank $(A) = r$.

i) If rank $(A + F) > r$ then $\|(A + F)^+\| \geq \|F\|^{-1}$.

ii) If rank $(A + F) \leq r$ and $\|A^+\| \cdot \|F\| < 1$ then rank $(A + F) = r$ and

$$\|(A + F)^+\| \leq \frac{\|A^+\|}{1 - \|A^+\| \, \|F\|} .$$

<u>Proof:</u> Let $\tau_1 \geq \ldots \geq \tau_n$ and $\sigma_1 \geq \ldots \geq \sigma_n$ be the singular values of $A + F$ and A respectively and let $s := $ rank $(A + F)$. Then $\tau_s > \tau_{s+1} = 0$, $\sigma_r > \sigma_{r+1} = 0$.

i) Since $s \geq r+1$ we obtain from Theorem 8.9

$$\|F\| = \|(A + F) - A\| \geq \tau_s - \sigma_s = \tau_s .$$

Corollary 8.4 vi) implies $\|(A + F)^+\| = \dfrac{1}{\tau_s} \geq \|F\|^{-1}.$

144

ii)　We have　$\|A^+\| = \dfrac{1}{\sigma_r}$,　$\|F\| < \sigma_r$　and　$\tau_r > 0$　by Theorem 8.9. Therefore　rank $(A + F) \geq r$　so that rank $(A + F) = r$. Again by Theorem 8.9

$$\| (A + F)^+\| = \frac{1}{\tau_r} \leq \frac{1}{\sigma_r - \|F\|} = \frac{\|A^+\|}{1 - \|A^+\| \, \|F\|} \quad .$$

$\square$

As we shall see in the next chapter the result ii) in Lemma 8.24 has serious consequences for the computation of pseudo-solutions.

<u>Lemma 8.25</u>

Let　$A, F \in \mathbb{R}^{m,n}$　and　$y, u \in \mathbb{R}^m$; let　$x := A^+ y$,
$\tilde{x} := (A + F)^+ (y + u)$, $d := y - Ax$. Then the following identity holds:

$$\tilde{x} - x = - (A + F)^+ Fx + (A + F)^+ \left((A + F)^+\right)^t F^t d$$

$$+ \, (I - (A + F)^+ (A + F))\, F^t (A^+)^t x + (A + F)^+ u.$$

<u>Proof:</u>　We have with　$B := A + F$:

$$\tilde{x} - x = (B^+ - A^+) y + B^+ u$$

$$= (- B^+ F A^+ + B^+ - A^+ + B^+ (B - A) A^+)\, y + B^+ u$$

$$= (- B^+ F A^+ + B^+ (I - AA^+) - (I - B^+ B) A^+)\, y + B^+ u \quad .$$

Since　$B^+ = B^+ B B^+ = B^+ (B^+)^t B^t$,

$$A^+ (I - AA^+) = A^t - A^t AA^+ = A^t - A^t (A^+)^t A^t$$

$$= A^t - A^t (A^t)^+ A^t = \Theta \quad ,$$

$$A^+ = A^+ (A^+)^t A^+ \quad , \quad (I - B^+ B) B^t = \Theta \quad ,$$

we obtain

$$B^+ (I - AA^+) = B^+ (B^+)^t F^t (I - AA^+) \quad ,$$

$$(I - B^+ B) A^+ = - (I - B^+ B) F^t (A^+)^t A^+ \quad .$$

From this we conclude

$$\tilde{x} - x = (- B^+FA^+ + B^+(B^+)^tF^t(I - AA^+) + (I - B^+B)F^t(A^+)^t A^+)y$$

$$+ B^+u = - B^+Fx + B^+(B^+)^tF^td + (I - B^+B)F^t(A^+)^tx$$

$$+ B^+u. \qquad \square$$

<u>Theorem 8.26</u>

Let $A,F \in \mathbb{R}^{m,n}$ and $y,u \in \mathbb{R}^m$; let $x := A^+y \neq \Theta$, $\tilde{x} := (A + F)^+(y + u)$ and $d := y - Ax$. If $\text{rank}(A) = \text{rank}(A + F)$ and $\|A^+\| \cdot \|F\| < 1$ then

$$(8.11) \qquad \frac{\|\tilde{x}-x\|}{\|x\|} \leq \frac{\|A\| \, \|A^+\|}{1-\|A^+\| \, \|F\|}.$$

$$\cdot \left\{ 2 \, \frac{\|F\|}{\|A\|} + \frac{\|A^+\| \, \|A\|}{1-\|A^+\| \, \|F\|} \cdot \frac{\|F\|}{\|A\|^2} \cdot \frac{\|d\|}{\|x\|} + \frac{\|u\|}{\|A\| \, \|x\|} \right\} .$$

<u>Proof:</u> By Lemma 8.25 and Lemma 8.24 we have

$$\|\tilde{x} - x\| \leq \| (A + F)^+\| \, \|F\| \, \|x\| + \| (A + F)^+\|^2 \|F\| \, \|d\|$$

$$+ \| I - (A + F)^+(A + F)\| \, \|F\| \, \|A^+\| \, \|x\|$$

$$+ \| (A + F)^+\| \, \|u\|$$

$$\leq \frac{\|A^+\| \, \|F\|}{1-\|A^+\| \, \|F\|} \|x\| + \frac{\|F\| \, \|A^+\|^2}{(1-\|A^+\| \, \|F\|)^2} \|d\|$$

$$+ \|F\| \, \|A^+\| \, \|x\| + \frac{\|A^+\|}{1-\|A^+\| \, \|F\|} \|u\|$$

$$= \frac{\|A^+\|}{1-\|A^+\| \, \|F\|} \left\{ \|F\| \cdot \|x\| + (1 - \|A^+\| \cdot \|F\|) \, \|F\| \cdot \|x\| + \right.$$

$$\left. + \frac{\|A^+\| \, \|F\|}{1-\|A^+\| \, \|F\|} \, \|d\| + \|u\| \right\}$$

which gives the result. $\qquad \square$

Definition 8.27

Let $A \in \mathbb{R}^{m,n}$. Then the number $\kappa(A) := \|A\| \; \|A^+\|$
is called the (generalized) <u>condition-number</u> of A.

It follows from the singular value decomposition of A that

$$\kappa(A) = \frac{\sigma_1}{\sigma_r}$$

where σ_1 and σ_r denote the largest and smallest non-zero
singular values of A respectively.

Remark 8.28

An interesting term in the upper bound in the estimate
(8.11) is the second summand

$$\frac{\kappa(A)^2}{(1-\|A^+\| \, \|F\|)^2} \quad \frac{\|F\|}{\|A\|^2} \; \cdot \; \frac{\|y-AA^+y\|}{\|A^+y\|} \quad .$$

It shows that $\kappa(A)^2$ measures the sensitivity of A^+y if
the residual $y - AA^+y$ is non-zero. One can construct an
example which shows that this dependence of the sensitivity
on $\kappa(A)^2$ is realistic.   *

Perturbation results may also proved in other norms. We consider as an illustration a case in which the Frobenius norm
is used.
Let us first prove an auxiliary result.

Lemma 8.29

Let P, Q be symmetric matrices in $\mathbb{R}^{k,k}$ with

$$P^2 = P \, , \; Q^2 = Q \, , \; \text{rank}(P) = \text{rank}(Q) \; .$$

Then

$$\|Q(I - P)\|_F = \|(I - Q)P\|_F \; .$$

147

<u>Proof:</u> Let $P' := I - P$, $Q' := I - Q$. Then $QQ' = Q'Q = \Theta$, $PP' = P'P = \Theta$.

We show first that $Q'P$ and QP' have the same singular values in $(0,1)$.

Let $\sigma \in (0,1)$ be a singular value of $Q'P$. Then

$$(Q'P)^t(Q'P)u = \sigma^2 u \quad \text{for some} \quad u \in \mathbb{R}^k \quad \text{with} \quad u \neq \Theta.$$

Simple consequences are

$$(*) \quad (PQ'P)u = \sigma^2 u \ , \ Pu = u \ , \ PQ'u = \sigma^2 u \ , \ Q'u \neq \Theta.$$

Since

$$\Theta = QQ'u = Q(P'Q'u + PQ'u) = QP'Q'u + QPQ'u$$

and

$$\Theta = P'Pu = P'u = P'(Qu + Q'u) = P'Qu + P'Q'u$$

we obtain with $(*)$

$$\Theta = QP'Q'u + \sigma^2 Qu$$

and

$$\Theta = P'QP'Q'u + \sigma^2 P'Qu = P'QP'Q'u - \sigma^2 P'Q'u \ .$$

Therefore if we set $v := P'Q'u$ we have

$$(QP')^t(QP')v = (P'QP')v = \sigma^2 v \ .$$

Since

$$Q'u = P'Q'u + PQ'u = v + \sigma^2 u$$

v cannot be zero because otherwise the identity

$$Q'u = Q'(Q'u) = \sigma^2 Q'u$$

would imply $Q'u = \Theta$ ($\sigma \in (0,1)$!) which contradicts $(*)$. This shows that σ is a singular value of QP' .

By the same arguments, each singular value of QP' in $(0,1)$ is also a singular value of $Q'P$. Next, by using the singular value decomposition it is easy to see that

$$(\#) \quad \|Q'P\|_F^2 = \sum_{i=1}^k \sigma_i(Q'P)^2 \ , \ \|QP'\|_F^2 = \sum_{i=1}^k \sigma_i(QP')^2 \ .$$

Since $Q'P$ and QP' have the same singular values in $(0,1)$ and since $Q'P$ and QP' have the same rank the result follows from $(\#)$.

In the proof of the following theorem we use repeatedly that

$$\|MN\|_F \leq \|M\| \ \|N\|_F$$

$$\|MN\|_F \leq \|M\|_F \ \|N\|$$

for all $M \in \mathbb{R}^{m,n}$, $N \in \mathbb{R}^{n,m}$.

Theorem 8.30

Let $A,F \in \mathbb{R}^{m,n}$ with $\operatorname{rank}(A + F) = \operatorname{rank}(A)$.
Then

$$\|(A + F)^+ - A^+\|_F \leq \|A^+\| \ \|(A + F)^+\| \ \|F\|_F .$$

__Proof:__ Let $B := A + F$ and

$$E_{11} := AA^+(A - B)B^+B = - A(A^+ - B^+)B ,$$

$$E_{12} := AA^+(A - B)(I - B^+B) = A(I - B^+B) ,$$

$$E_{21} := (I - AA^+)(A - B)B^+B = - (I - AA^+)B ,$$

$$F_{11} := A^+A(A^+ - B^+)BB^+ = - A^+(A - B)B^+ ,$$

$$F_{12} := A^+A(A^+ - B^+)(I - B^+B) = A^+(I - BB^+) ,$$

$$F_{21} := (I - A^+A)(A^+ - B^+)BB^+ = - (I - A^+A)B^+ .$$

Then, using repeatedly that $\|U + V\|_F^2 = \|U\|_F^2 + \|V\|_F^2$ if
$U^t V = \theta$ or $V^t U = \theta$, it follows

$$\|A - B\|_F^2 = \sum_{i,j} \|E_{ij}\|_F^2 , \quad \|A^+ - B^+\|_F^2 = \sum_{i,j} \|F_{ij}\|_F^2 .$$

We have

$$F_{11} = - A^+E_{11}B^+ , \quad \|F_{11}\|_F \leq \|A^+\| \ \|E_{11}B^+\|_F$$

$$\leq \|A^+\| \ \|E_{11}\| \ \|B^+\|_F.$$

From Lemma 8.29 follows

$$\|AF_{12}\|_F \;=\; \|AA^+(I - B^+B)\|_F \;=\; \|(I - AA^+)BB^+\|_F$$

$$=\; \|E_{21}B^+\|_F \;\le\; \|E_{21}\|_F\,\|B^+\| \;,$$

$$\|F_{21}B\|_F \;=\; \|(I - A^+A)B^+B\|_F \;=\; \|A^+A(I - B^+B)\|_F$$

$$=\; \|A^+E_{12}\|_F \;\le\; \|E_{12}\|_F\,\|A^+\| \;.$$

Hence

$$\|F_{12}\|_F \;=\; \|A^+AF_{12}\|_F \;\le\; \|A^+\|\,\|B^+\|\,\|E_{21}\|_F \;,$$

$$\|F_{21}\|_F \;=\; \|F_{21}BB^+\|_F \;\le\; \|E_{12}\|_F\,\|A^+\|\,\|B^+\| \;.$$

Since $\;\|A - B\|_F^2 = \|E_{11}\|_F^2 + \|E_{12}\|_F^2 + \|E_{21}\|_F^2\;$ the result follows from the estimates above. $\qquad\square$

8.5 An application: Fitting of data

The least squares fit of experimental data is a common tool in many applied sciences. Suppose that we wish to fit a function of the form

$$z(t;a,b) \;:=\; \sum_{j=1}^{n} a_j\, \eta_j(t;b)$$

to a set of data (t_i, z_i) , $1 \le i \le m$; here the functions $\eta_j(.;b)$ are known functions while $b = (b_1, \ldots, b_k)$ and $a = (a_1, \ldots, a_n)$ are parameters to be estimated. We define the "best" fit to the data by minimizing the (nonlinear) functional

$$r(a,b) \;:=\; \sum_{i=1}^{m} (z_i - z(t_i;a,b))^2$$

with respect to $a \in \mathbb{R}^n$ and $b \in \Omega \subset \mathbb{R}^k$.

Examples 8.31

1) Polynomial fitting: $z(t;a,b) = z(t;a) = \sum_{j=1}^{n} a_j\, t^{j-1}$.

$$2) \quad z(t;a,b) := a_1 \cos(b_1 t) + a_2 \sin(b_2 t) + a_3 \ .$$

A family of such functions is used if the data are derived from a periodic process.

$$3) \quad z(t;a,b) := \sum_{j=1}^{n} a_j e^{b_j t} \ .$$

Many situations in biology, economics and the physical sciences yield data which can best be described by the family above.   *

Let us define

$$\Phi(b) := (\varphi_{ij}(b)), \quad \varphi_{ij}(b) := \eta_j(t_i;b), \quad 1 \leq i \leq m, \ 1 \leq j \leq n,$$

$$z := (z_1,\ldots,z_m) \in \mathbb{R}^m \ .$$

Then we can rewrite $r(a,b)$ as

$$r(a,b) = \| z - \phi(b)a \|^2 \ .$$

One approach to solve the minimization problem

$$\text{minimize} \quad r(a,b) \quad \text{subject to} \quad a \in \mathbb{R}^n \ , \ b \in \Omega$$

consists in separating the linear variables a and the non-linear parameter b . For any value of b, we minimize $r(.;b)$ by choosing $a_b := \phi(b)^+ z$. Substituting a_b into r gives

$$e(b) := r(a_b,b) = \| z - \phi(b)\phi(b)^+ z \|^2 \ .$$

Since $I - \phi(b)\phi(b)^+$ is the orthogonal projector onto $R(\phi(b))$ to minimize e with respect to $b \in \Omega$ it is equivalent to maximize

$$\hat{e}(b) := \| \phi(b)\phi(b)^+ z \|^2$$

over Ω . If $\hat{b}$ is a best parameter with respect to $\hat{e}$ (or e) we define a "best" parameter $\hat{a}$ by defining

$$\hat{a} := \phi(\hat{b})^+ z \ .$$

Now, a-priori it is not clear if this pair $(\hat{a},\hat{b})$ is also a minimizing vector of the original objective function r .

We don't give here details of the analysis of this approach.
The main problem results from the fact that the mapping

$$b \longmapsto \| \phi(b)\phi(b)^{+}z \|^{2}$$

is difficult to handle theoretically and numerically.

Bibliographical comments

The material presented in this chapter can be found for the
most part in any book on generalized inverses; the reader is
referred to BEN-ISRAEL, GREVILLE [11] and CAMPELL, MEYER [17].
The weighted pseudo-inverse which is used to solve least
squares problems with linear constraints is described in
ELDÉN [29] and LÖTSTEDT [67]. Various sharper results on the
dependence of pseudo-solutions on perturbations of the data
can be found in WEDIN [101] and v.d. SLUIS, VELTKAMP [96]. For
a complete discussion of the fitting problem the reader is referred to GOLUB, PEREYRA [39].

Chapter 9: Numerical aspects of least squares problems

This chapter presents and analyses some problems which are related to the computation of least squares solutions. The aim is to make the reader familiar with the main tools and methods to calculate such solutions.

9.1 Calculation of A^+: The factorization approach.

Let us assume here and throughout this chapter that A is a matrix in $\mathbb{R}^{m,n}$, $m \geq n$, with rank r and that y is a vector in $\mathbb{R}^m$.

If A has full rank, that is if $r = n$, then we know that $A^+ = (A^tA)^{-1}A^t$. The classical approach to compute the pseudo-solution A^+y in this case is via the system of normal equation

$$(9.1) \qquad A^tAx = A^ty.$$

The equation (9.1) can efficiently be solved by factorizing A^tA in the following form

$$(9.2) \qquad A^tA = U^tU$$

where U is an upper triangular nonsingular matrix in $\mathbb{R}^{n,n}$.

The factorization in the form (9.2) can be achieved by Cholesky's method; the upper triangular matrix U is usually called the Cholesky-factor of A^tA. If we have the decomposition (9.2) then the pseudo-solution A^+y can be obtained by solving

$$(9.3) \qquad U^tz = A^ty \ , \ Ux = z.$$

Notice that the equations in (9.3) can be solved by forward substitution and by back substitution.

Unfortunately, the normal equations method has an important drawback:

There may be a serious loss of information in explicitly forming and processing $A^t A$ and $A^t y$.

The following example illustrates this statement:

Illustration 9.1

If $A = \begin{pmatrix} 1 & 1 \\ \delta & 0 \\ 0 & \delta \end{pmatrix}$ then $A^t A = \begin{pmatrix} 1+\delta^2 & 1 \\ 1 & 1+\delta^2 \end{pmatrix}$.

If $A^t A$ is computed by a computer with accuracy 10^{-t} then $A^t A$ is singular for every δ with $|\delta| \leq 10^{-\frac{t}{2}}$.   *

A better way to compute $A^+ y$ if A has full rank is to use a bidiagonalization of A. It can be shown that there are orthogonal matrices $U \in \mathbb{R}^{m,n}$ and $V \in \mathbb{R}^{n,n}$ such that

$$(9.4) \qquad U^t A V = \begin{pmatrix} B \\ \Theta \end{pmatrix}$$

where B is nonsingular and bidiagonal:

$$B = \begin{pmatrix} \alpha_1 & \beta_2 & & & \\ & \cdot & \cdot & & \Theta \\ & & \cdot & \cdot & \\ & & & \cdot & \beta_n \\ \Theta & & & & \alpha_n \end{pmatrix}$$

The algorithm to achieve this decomposition is based on Householder transformations.

Definition 9.2

Let $w \in \mathbb{R}^k$ with $\|w\| = 1$. Then the matrix H of the form

$$H_w = I - 2 w w^t$$

is called a __Householder matrix__.

Householder matrices are symmetric and orthogonal. They are used to zero specified entries in a matrix or a vector:

> Given $z \in \mathbb{R}^k$ with $z \notin \text{span}(e^1)$, then the Householder matrix

$$(9.5) \qquad H_w := I - 2ww^t \text{ where}$$

$$w := (z + \|z\| e^1) \, \|z + \|z\| e^1\|^{-1}$$

> has the property that $H_w z = \|z\| e^1$.

Using these matrices it is obvious how to introduce zeros into rows and columns of a matrix.

If A is transformed into the bidiagonal form (9.4) the solution $x = A^+ y$ is obtained by solving the following problems

$$Bz = u \ , \ x = Vz$$
$$(9.6)$$
$$\text{where } U^t y = \begin{pmatrix} u \\ v \end{pmatrix} \ , \ u \in \mathbb{R}^n \ , \ v \in \mathbb{R}^{m-n}.$$

Since B is a bidiagonal matrix the computation of z in (9.6) is very simple.

Now we come to the case that A is rank-deficient, that is if $r < n$. The tactics also in this case is to convert the problem of computing $A^+ y$ into a sequence of easy - to - solve problems.

The following lemma contains the main idea:

Lemma 9.3

Let
$$(9.7) \qquad A = BC \text{ where } B \in \mathbb{R}^{n,r} \ , \ C \in \mathbb{R}^{r,n} \text{ with}$$

$$\text{rank}(B) = \text{rank}(C) = r.$$

Then

$$A^+ = C^+ B^+ \ , \ C^+ = C^t (CC^t)^{-1} \ , \ B^+ = (B^t B)^{-1} B^t.$$

Proof: Since the matrices B and C have full rank the representation of B^+ and C^+ are given in the form above. We verify that $G := C^+B^+$ satisfies the properties c) in Theorem 8.13 by using the identities $B^+B = I$, $CC^+ = I$ and the properties in c) of Theorem 8.13 for B and C:

$$AC^+B^+A = BCC^+B^+BC = BC = A \; ,$$
$$C^+B^+AC^+B^+ = C^+B^+BCC^+B^+ = C^+B^+ \; ,$$
$$(AC^+B^+)^t = (BCC^+B^+)^t = (BB^+)^t = BB^+ = BCC^+B^+ = AC^+B^+ \; ,$$
$$(C^+B^+A)^t = (C^+B^+BC)^t = (C^+C)^t = C^+C = C^+B^+BC = C^+B^+A.$$

$\square$

If we have a factorization of the form (9.7) then the problem of computing $x = A^+y$ is reduced to a sequence of (hopefully simple solvable) problems

$$z = B^+y \; , \quad x = C^+z.$$

The various methods to compute the solution $x := A^+y$ are distinguished by the way how A is factorized in the form (9.7). In the following we sketch three methods. Let us begin with the

LU-decomposition

(9.8) $A = P^tLUQ^t$

where $P \in {\rm I\!R}^{m,m}$, $Q \in {\rm I\!R}^{n,n}$, $L = (l_{ij}) \in {\rm I\!R}^{m,r}$,

$U = (u_{ij}) \in {\rm I\!R}^{r,n}$ such that

$l_{ii} = 1, \quad 1 \le i \le r, \quad l_{ij} = 0 \; , \; i < j \; ,$

$u_{ii} \ne 0, \quad 1 \le i \le r, \quad u_{ij} = 0 \; , \; i > j \; ,$

Q,P are permutation matrices.

It is well-known that this factorization can be achieved by both row and column interchanges ("complete pivoting") and Gaussian elimination. Complete pivoting is very important since a situation as given in Ex.9.4 is quite possible when solving a least squares problem. The pseudo-solution A^+y is obtained from (9.8) by solving the sequence

(9.9) $u = L^t Py$, $L^t Lv = u$, $UU^t w = v$, $x = QU^t w$.

Each of $L^t L$ and UU^t is a positive definite matrix in $\mathbb{R}^{r,r}$ and can be factorized by Cholesky's method.

Example 9.4

If
$$A = \begin{pmatrix} \delta & 0 & 0 \\ 1 & 0 & 1 \\ 0 & 1 & 2 \\ 1 & 1 & 2 \end{pmatrix}$$

then A has full rank for every δ but if $|\delta|$ is small we cannot use the element δ as a pivot. (In solving linear equations in the ordinary sense a row with only one small entry would imply that the matrix is singular).   *

The QR-decomposition

(9.10) $A = QRP^t$

where $R = (r_{ij}) \in \mathbb{R}^{r,n}$, $Q \in \mathbb{R}^{m,r}$, $P \in \mathbb{R}^{n,n}$

such that

   i)   R is upper triangular ($r_{ij} = 0$ for $i > j$);

   ii) $r_{ii}^2 \geq \sum\limits_{k=i}^{r} r_{kj}^2$, $i + 1 \leq j \leq n$, $1 \leq i \leq r$;

   iii) P is a permutation matrix ;

   iv) $Q^t Q = I$.

The matrix Q is usually obtained as a product of Householder or Givens transformations or by Gram-Schmidt orthogonalization. The permutation matrix P is chosen by the method of column pivoting which implies the property ii) in (9.10). Notice that the matrix R is a Cholesky factor of A if $P = I$.

The pseudo-solution $x = A^+ y$ is obtained from (9.10) by solving the sequence

(9.11) $u = Q^t y$, $Rz = u$, $x = Pz$.

The singular value decomposition

(9.12) $A = UDV^t$

where $U \in \mathbb{R}^{m,r}$, $D = \mathrm{diag}(\sigma_1,\dots,\sigma_r) \in \mathbb{R}^{r,r}$,
$V \in \mathbb{R}^{r,n}$ such that

i) $U^t U = I$, $VV^t = I$,

ii) $\sigma_1 \geq \dots \geq \sigma_r > 0$.

The computation of the singular value decomposition is usually organized in two phases:

Phase 1: Transformation of A into a bidiagonal matrix B by Householder transformations; the singular values of B are the same as those of A.

Phase 2: Computation of the singular value decomposition of the bidiagonal matrix B by zeroing the superdiagonal elements in B. This is done by an iterative process.

It is important that the method above avoids the explicit computation of $A^t A$ which would be necessary if the definition of the singular values would be used in a naive way. The pseudo-solution $A^+ y$ is obtained from (9.12) by solving the sequence

(9.13) $u = D^{-1} U^t y$, $x = Vu$.

Remarks 9.5

1) In general, the matrix A and the data y are transformed simultaneously with considerable savings in computations.

2) If one is only interested in a least squares solution the amount of computational work can be reduced.

3) Notice that in solving a least squares problem it is not possible to scale the rows of the matrix without changing the least squares objective function.

*

9.2 Rank decision

In the last section we considered the problem of computing A^+ from a decomposition

$$(9.14) \qquad A = BC \; , \quad B \in {\rm I\!R}^{m,r} \; , \quad C \in {\rm I\!R}^{r,n}$$

$$\text{where } r = \text{rank}\,(A) = \text{rank}(B) = \text{rank}\,(C) \; .$$

If we compute the decomposition (9.14) under the influence of rounding errors we obtain

$$(9.15) \qquad \bar{A} := \overline{BC} = A + F \; , \quad \bar{B} \in {\rm I\!R}^{m,q} \; , \quad \bar{C} \in {\rm I\!R}^{q,n}$$

$$\text{rank}\,(\bar{B}) = \text{rank}\,(\bar{C}) = q$$

where $\bar{A}$, $\bar{B}$, $\bar{C}$ are the computed quantities and F is the matrix which describes the effect of rounding. As a consequence

$$(9.16) \qquad \bar{A}^+ = \bar{C}^+ \, \bar{B}^+.$$

In general (due to rounding errors) we may expect $q > r$ if A is rank-deficient, that is if $r = \text{rank}(A) < n$. As we know from L.8.24 the pseudo-inverse $\bar{A}^+$ may be very different from A^+ if $q > r$. Thus, we have to apply a so called rank decision criterion to get back the rank of A which we don't know before. The principal problem is that we must work with the quantities $\bar{B}$, $\bar{C}$ in (9.15) not with B,C in (9.14).

Definition 9.6

The number

$$\Psi_h(A;A + F) := \inf\{\text{rank}(A + F + F') \,\big|\, \|F'\| \le h\}$$

is called the __pseudo-rank of A at level__ $h > 0$

__based on__ $A + F$.

Clearly, $\Psi_h(A;A + F) \le \text{rank}(A + F)$ for every $h \ge 0$. According to L.8.24 it must be the aim to choose the level h such that $\Psi_h(A;A + F) \le \text{rank}(A)$. Then if $\|F\| + h < \|A^+\|^{-1}$ and F' is any matrix with $\|F'\| \le h$ and $\text{rank}(A + F + F') = \Psi_h(A;A + F)$ we know that $(A + F + F')^+$ may be a good approximation for A^+.

However the choice of the level h is a very difficult problem.
We are (again) in the situation to choose an trade-off-para-
meter:

- If h is too large then $A + F + F'$ may be a bad
  approximation for A.

- If h is too small then the pseudo-rank Ψ_h may be larger
  then the rank of A which implies that $(A + F + F')^+$ is a
  bad approximation for A^+.

Of course, an information on the size of $\|F\|$ is very useful
in finding out the best level h. An upper bound for $\|F\|$ must
be derived from a detailed error analysis. For the usual factorization methods of the form (9.14) an estimation of the following form is known:

$$(9.17) \qquad \|F\| \leq c\rho \|\overline{B}\| \, \|\overline{C}\|$$

where ρ is the rounding unit for the arithmetic in question
and c is a small constant depending on the numbers m,n and q.

Let us consider the three factorization methods used in
the last section.

<u>The singular value decomposition</u>

Let

$$A + F = \overline{U}\,\overline{D}\,\overline{V}^t \, , \quad \overline{U} \in \mathbb{R}^{m,q} \, , \quad \overline{V} \in \mathbb{R}^{q,n} \, ,$$

$$D = \mathrm{diag}(\overline{\sigma}_1,\ldots,\overline{\sigma}_q) \in \mathbb{R}^{q,q} \, , \quad \overline{\sigma}_1 \geq \ldots \geq \overline{\sigma}_q \, .$$

A reasonable criterion for the rank-decision is given by the
following rule $(\overline{\sigma}_{q+1} := 0)$:

Choose $p \leq q$ such that

$$(9.18) \qquad \overline{\sigma}_{p+1} \leq h < \overline{\sigma}_p \, .$$

If p is chosen according to (9.18) then the pseudo-rank
$\Psi_h(A;A + F)$ is p. This can be seen as follows: If $F' := \overline{U}D'\overline{V}^t$
with $D' = -\mathrm{diag}(0,\ldots,0,\, \overline{\sigma}_{p+1},\ldots,\overline{\sigma}_q)$ then $\|F'\| \leq \|D'\| = \overline{\sigma}_{p+1} \leq h$
and therefore $\Psi_h(A;A + F) \leq p$. On the other hand, if F' is any

matrix with $\|F'\| \leq h$ then $\sigma_p(A + F + F') \geq \sigma_p(A + F) - \|F'\|$ $\geq \bar{\sigma}_p - h > 0$ and therefore rank $(A + F + F') \geq p$. To choose the tolerance parameter h an upper bound for $\|F\|$ is very helpful since no singular value of $A + F$ corresponding to a zero singular value of A can be larger than $\|F\|$.

The QR-decomposition

Let
$$A + F = \bar{Q}\,\bar{R}\,\bar{P}^t, \quad \bar{Q} \in \mathbb{R}^{m,q}, \quad \bar{P} \in \mathbb{R}^{n,n},$$
$$\bar{R} = (\bar{r}_{ij}) \in \mathbb{R}^{q,n} \text{ upper triangular,}$$
$$\bar{r}_{ii}^2 \geq \sum_{k=i}^{q} \bar{r}_{kj}^2, \quad i + 1 \leq j \leq n, \ 1 \leq i \leq q.$$

A reasonable rank-decision criterion is given as follows $(\bar{r}_{q+1,q+1} := 0)$:

Choose $p \leq q$ such that

(9.19) $|\bar{r}_{p+1,p+1}| \leq h < \bar{r}_{p,p}$.

Let
$$\bar{R} = \begin{pmatrix} \bar{R}_{11} & \bar{R}_{12} \\ \Theta & \bar{R}_{22} \end{pmatrix}, \quad \bar{R}_{11} \in \mathbb{R}^{p,p} .$$

Then the pseudo-rank $\Psi_h(A; A + F)$ is not larger then p. This follows from the fact that $F' := \bar{Q}R'\bar{P}^t$ with $R' = \begin{pmatrix} \Theta & \Theta \\ \Theta & R_{22} \end{pmatrix}$ satisfies $\|F'\| \leq h$ due to the pivot strategy and the choice of p.

The LU-decomposition

Let
$$A + F = \bar{P}^t\,\bar{L}\,\bar{U}\,\bar{Q}^t, \quad \bar{P} \in \mathbb{R}^{m,m}, \quad \bar{Q} \in \mathbb{R}^{n,n}, \quad \bar{L} \in \mathbb{R}^{m,q},$$
$$\bar{U} = (\bar{u}_{ij}) \in \mathbb{R}^{q,n} \text{ upper trian-}$$
$$\text{gular.}$$

A rank-decision criterion is given by the following rule $(\bar{u}_{q+1,q+1} := 0)$:

Choose $p \leq q$ such that

(9.20) $\qquad |\bar{u}_{p+1,p+1}| \leq h < |\bar{u}_{p,p}|$.

Remark 9.7

In practice, factorization is a dynamic process: Starting with $A^{(o)} := A$ a sequence $A^{(1)},...,A^{(q)} =: A + F$ is computed. This leads to the fact that the rank-decision process is a dynamic process too: At each factorization step the rank-decision criterion has to be applied. $\qquad\qquad *$

9.3 Cross-validation

As we know from Chapter 6 we can approximate A^+ by the damped pseudo-inverse A_α^{-1} given by

$$A_\alpha^{-1} := (A^t A + \alpha I)^{-1} A^t$$

where the parameter $\alpha > o$ has to be chosen properly. In Chapter 6 there are also given rules how to choose this trade-off parameter α. In this section we consider an a-posteriori strategy for choosing α which doesn't use any information on the size of the noise level.

Let $a_1,...,a_m$ be the rows of the matrix A and let $y_1,...,y_m$ be the components of the vector y. Then the original equation is given by

(9.21) $\qquad a_i^t x = y_i$, $1 \leq i \leq m$.

The idea of cross-validation as a strategy for choosing α is the following one: Let $x^{\alpha,k}$ be a solution of

$$\min\{\frac{1}{m} \sum_{\substack{i=1 \\ i \neq k}}^{m} (a_i^t x - y_i)^2 + \alpha \|x\|^2 \mid x \in \mathbb{R}^n \}.$$

Thus, $x^{\alpha,k}$ is the Tikhonov regularized solution of (9.21) with the k-th row deleted. In a practical situation this is equivalent with the ignoration of the k-th experiment.

For each $\alpha > o$ we measure the closeness of $a_k^t x^{\alpha,k}$ to y_k by the weighted mean square error

$$V(\alpha) \; := \; \frac{1}{m} \sum_{k=1}^{m} \; (a_k^t \, x^{\alpha,k} - y_k)^2 \, w_k(\alpha) \; , \alpha > o \, ,$$

where the weights w_k are to be chosen appropriate. Let

$$G(\alpha) \; := \; (g_{ij}(\alpha)) \; := \; A(A^tA + \alpha I)^{-1}A^t.$$

From statistical arguments it is reasonable to choose

$$w_k(\alpha) \; := \; \{(1 - g_{kk}(\alpha))(\frac{1}{m} \, tr(I - G(\alpha))^{-1})\}^2.$$

With this choice we arrive at

$$V(\alpha) = \frac{1}{m} \, \|(I - G(\alpha))y\|^2 (\frac{1}{m} \, tr(I - G(\alpha)))^{-2}.$$

(9.22) $$= m\|y - A \, x^{\alpha}\|^2 \, tr(I - G(\alpha))^{-2}$$

where $x^{\alpha} = A_{\alpha}^{-1}y$. Therefore, the criterion to choose α consits in solving the following problem

(9.23)   Minimize $V(\alpha)$ subject to $\alpha > o$.

To solve this minimization problem we have to compute function values $V(\alpha)$. This can be done in an efficient way if we have at hand a bidiagonalization of A (see (9.4)):

$$A = U\binom{B}{\Theta}V^t$$

where B is bidiagonal and $U \in \mathbb{R}^{m,m}$, $V \in \mathbb{R}^{n,n}$ are orthogonal matrices. Then if

$$\binom{u}{v} = U^t y$$

the computation $x^{\alpha} := A_{\alpha}^{-1}y$ is equivalent via the transformation $x = Vz$ to

(9.24) $\min \{ \|Bz - u\|^2 + \alpha\|z\|^2 \mid z \in \mathbb{R}^n \}$

or

(9.25) $\min \{ \|\binom{B}{\sqrt{\alpha}I} z - \binom{u}{\Theta}\|^2 \mid z \in \mathbb{R}^n \}.$

The second minimization problem is efficiently solved by computing a QR-decomposition of $\binom{B}{\sqrt{\alpha}I}$ (see Section 9.1):

$$(9.26) \qquad Q^t \begin{pmatrix} B \\ \sqrt{\alpha}I & \Theta \end{pmatrix} \begin{pmatrix} u \\ \end{pmatrix} = \begin{pmatrix} B(\alpha) & w_1 \\ \Theta & w_2 \end{pmatrix} ;$$

here $B(\alpha)$ is a bidiagonal matrix:

$$(9.27) \qquad B(\alpha) = \begin{pmatrix} c_1 & d_1 & & & \\ & \cdot & \cdot & & \Theta \\ & & \cdot & \cdot & \\ & & & \cdot & d_{n-1} \\ \Theta & & & & c_n \end{pmatrix} .$$

The solution z^α of (9.25) is then obtained by solving $B(\alpha)z = w_1$. Now we obtain

$$\begin{aligned}
\|Ax^\alpha - y\|^2 &= \|AVz^\alpha - y\|^2 \\
&= \left\| U\begin{pmatrix} B \\ \Theta \end{pmatrix} V^t V z^\alpha - U\begin{pmatrix} u \\ v \end{pmatrix} \right\|^2 \\
&= \|Bz^\alpha - u\|^2 + \|v\|^2 .
\end{aligned}$$

The computation of the trace-term in (9.23) starts from

$$\begin{aligned}
\mathrm{tr}(I - G(\alpha)) = m - \mathrm{tr}(G(\alpha)) &= m - \mathrm{tr}(A(A^tA + \alpha I)^{-1}A^t) \\
&= m - \mathrm{tr}(I - \alpha(A^tA + \alpha I)^{-1}) \\
&= m - n + \alpha\,\mathrm{tr}((A^tA + \alpha I)^{-1}) \\
&= m - n + \alpha\,\mathrm{tr}((VB^tBV^t + \alpha I)^{-1}) \\
&= m - n + \alpha\,\mathrm{tr}((B^tB + \alpha I)^{-1}) .
\end{aligned}$$

Using the QR-decomposition (9.26) we see

$$B^tB + \alpha I = B(\alpha)^t B(\alpha)$$
$$\mathrm{tr}((B^tB + \alpha I)^{-1}) = \mathrm{tr}((B(\alpha)^t B(\alpha))^{-1}) = \sum_{i=1}^{m} \|b_i(\alpha)\|^2$$

where $b_1(\alpha),\ldots,b_n(\alpha)$ are the rows of $B(\alpha)^{-1}$.

Using the identity $B(\alpha)B(\alpha)^{-1} = I$ and the representation (9.27).
We obtain the identities

$$c_n b_n(\alpha) = e_n, \quad c_i b_i(\alpha) = e_i - d_i b_{i+1}(\alpha), \quad 1 \leq i \leq n - 1.$$

Since $B(\alpha)^{-1}$ is upper triangular b_{i+1} is orthogonal to e_i.

Hence

$$\| b_n(\alpha) \|^2 = c_n^{-2} \, ,$$

$$\| b_i(\alpha) \|^2 = (1 + d_i \| b_{i+1}(\alpha) \|^2) c_i^{-2} \, , \quad 1 \le i \le n - 1.$$

This gives us a recursion to compute the trace-term in $O(n)$ operations.

9.4 Successive Approximation

If we want to compute $A^+ y$ we can do this by solving the normal equation

$$A^t A x = A^t y$$

in $N(A)^\perp = R(A^t)$. The well known method of successive approximation defines the following iteration:

Choose $x_o \in \mathbb{R}^n$ and $\alpha > o$, continue with

(9.28)

$$x_{k+1} := x_k + \alpha A^t (y - A x_k).$$

Theorem 9.8

If $x_o \in R(A^t)$ and $\alpha < 2 \| A^t A \|^{-1}$ then the sequence $(x_k)_{k \in \mathbb{N}}$ generated by (9.28) converges to $A^+ y$.

Proof: Since $x_o \in R(A^t)$ we obtain $x_k \in R(A^t)$ for all $k \in \mathbb{N}$. Therefore if we can prove the convergence of the sequence the limit must be $A^+ y$. But the iteration is defined by the fixed point iteration

$$x_{k+1} := F(x_k)$$

where $F(x) := (I - \alpha A^t A) x + \alpha A^t y$. Since F is Lipschitz − continuous with Lipschitz-constant $L := \| I - \alpha A^t A \| < 1$ the result follows from Banach's fixed point theorem.

$\square$

Remarks 9.9

1) Banach's fixed point theorem shows that the convergence in Theorem 9.8 is of geometric type.

2) The iteration (9.28) may also be considered as a gradient method to minimize the objective functional $F(x) := \|Ax - y\|^2$; the stepsize $\alpha := \|A^tA\|^{-1}$ is a choice which is well-known in optimization theory.   *

In the following we analyse the iteration (9.28) under the influence of errors. Let $x^o \in \mathbb{R}^n$, $y^\varepsilon, y^o \in \mathbb{R}^m$ with

$$Ax^o = y^o, \quad \|y^\varepsilon - y^o\| \le \varepsilon.$$

If we apply Th.9.8 for $y := y^\varepsilon$ we obtain under the hypotheses of the theorem convergence to A^+y^ε. Since a precise realization of (9.28) is practically impossible the determination of each successive approximation involves errors. Therefore the procedure (9.28) for $y := y^\varepsilon$ with $\alpha = 1$ is actually

(9.29)
$$\text{Choose } z_o \in \mathbb{R}^n, \text{ continue with}$$
$$z_{k+1} := z_k + A^t(y^\varepsilon - Az_k) + u_k, \quad k \in \mathbb{N} \cup \{0\};$$

here u_k are the computational errors. We assume throughout in the following

(9.30)
a) $\|u_k\| \le \beta$, $k \in \mathbb{N}$;
b) $\|A\| \le 1$, $\|I - A^tA\| \le 1$, $\left\| (I - A^tA)\Big|_{R(A^t)} \right\| < 1$.

The sequence $(z_k)_{k \in \mathbb{N}}$ does not converge to x^o necessarily. Therefore, the iteration (9.29) must be supplemented by a stopping rule:

(9.31)
Given a tolerance number $\tau > o$, the stopping index $h = h(\tau)$ is determined by the following criterion:

$$\|z_{h+1} - z_h\| \le \tau, \quad \|z_{k+1} - z_k\| > \tau \text{ for all } k < h.$$

The problem consists in finding the tolerance level $\tau = \tau(\varepsilon, \beta)$ such that

$$\lim_{\varepsilon \to 0, \beta \to 0} z_{h(\tau(\varepsilon,\beta))} = x^{\circ}.$$

Before we prove such a result we have to prepare some arguments.

Let $B := I - A^t A$ and consider the iteration

$$(9.32) \qquad \begin{aligned} w_{\circ} &:= z_{\circ} \\ w_{k+1} &:= w_k + A^t(y^{\circ} - Aw_k) + u_k , \quad k \in \mathbb{N} \cup \{0\}. \end{aligned}$$

<u>Lemma 9.10</u>

We have:

i) $\displaystyle\sum_{j=1}^{k} \|w_j - w_{j+1} + u_j\|^2 \leq \|w_{\circ} - x^{\circ}\|^2 + \sum_{j=0}^{k-1} \|u_j\|^2$, $k \in \mathbb{N}$.

ii) $\displaystyle\varlimsup_{k} \|w_k - w_{k+1}\| \leq 2\beta$.

iii) $\displaystyle\varlimsup_{k} \|z_k - z_{k+1}\| \leq 2\beta$.

iv) $z_k - z_{k+1} = w_k - w_{k+1} - B^k A^t(y^{\varepsilon} - y^{\circ})$.

<u>Proof:</u> Let $\Delta_k := w_k - x^{\circ}$. Then $\Delta_{k+1} = B\Delta_k + u_k$ and

$$\sum_{j=0}^{k-1} \|u_j\|^2 = \sum_{j=0}^{k-1} \|\Delta_{j+1}\|^2 + \sum_{j=0}^{k-1} \|B\Delta_j\|^2 - 2 \sum_{j=0}^{k-1} (\Delta_{j+1}, B\Delta_j)$$

$$= \sum_{j=0}^{k-1} \|\Delta_{j+1}\|^2 + \sum_{j=0}^{k-1} \|B\Delta_j\|^2 - 2 \sum_{j=0}^{k-1} (B^{1/2}\Delta_{j+1}, B^{1/2}\Delta_j)$$

$$\geq \sum_{j=0}^{k-1} \|\Delta_{j+1}\|^2 + \sum_{j=0}^{k-1} \|B\Delta_j\|^2$$

$$- 2 \left(\sum_{j=0}^{k-1} (B\Delta_{j+1}, \Delta_{j+1}) \right)^{1/2} \left(\sum_{j=0}^{k-1} (B\Delta_j, \Delta_j) \right)^{1/2}$$

$$\geq -\|\Delta_o\|^2 + \sum_{j=o}^{k} ((I-B)\Delta_j, (I-B)\Delta_j)$$

$$+ 2 \sum_{j=o}^{k} (B\Delta_j, \Delta_j) - (B\Delta_k, B\Delta_k) - 2\gamma_k$$

where

$$\gamma_k := \{ \sum_{j=1}^{k} (B\Delta_j, \Delta_j) \}^{1/2} \{ \sum_{j=o}^{k-1} (B\Delta_j, \Delta_j) \}^{1/2}.$$

Since by assumption (9.30)

$$- (B\Delta_k, B\Delta_k) = - \|B\Delta_k\|^2 \geq - \|B^{1/2}\Delta_k\|^2 = - (B\Delta_k, \Delta_k)$$

and

$$\{ \sum_{j=o}^{k} (B\Delta_j, \Delta_j) \}^2 - (B\Delta_k, B\Delta_k) \{ \sum_{j=o}^{k} (B\Delta_j, \Delta_j) \} + \frac{1}{4}(B\Delta_k, B\Delta_k)^2$$

$$\geq \{ \sum_{j=1}^{k} (B\Delta_j, \Delta_j) \} \{ \sum_{j=o}^{k-1} (B\Delta_j, \Delta_j) \}$$

$$+ (B\Delta_k, \Delta_k) \sum_{j=o}^{k} (B\Delta_j, \Delta_j) - (B\Delta_k, B\Delta_k) \sum_{j=o}^{k} (B\Delta_j, \Delta_j)$$

we have

$$\sum_{j=o}^{k-1} \|u_j\|^2 \geq -\|\Delta_o\|^2 + \sum_{j=o}^{k} ((I-B)\Delta_j, (I-B)\Delta_j).$$

This proves part i).

Since

$$\varliminf_{k} \|w_k - w_{k+1}\| \leq \varliminf_{k} (\|w_k - w_{k+1} + u_k\| + \|u_k\|)$$

$$\leq \varlimsup_{k} \{\frac{1}{k} \sum_{j=o}^{k-1} \|w_j - w_{j+1} + u_j\|^2\}^{1/2} + \beta$$

$$\leq \varlimsup_{k} \{\frac{1}{k} \sum_{j=o}^{k-1} \|u_j\|^2\}^{1/2} + \beta \leq 2\beta$$

the assertion in ii) holds. From

$$z_k - z_{k+1} - (w_k - w_{k+1}) = A^t A(z_k - w_k) - A^t(y^\varepsilon - y^o)$$

and

$$z_o - w_o = 0, \quad z_k - w_k = \sum_{j=o}^{k-1} B^j A^t (y^\varepsilon - y^o), \quad k \geq 1,$$

we obtain

$$z_k - z_{k+1} - (w_k - w_{k+1}) = - B^k A^t (y^o - y^\varepsilon)$$

which is part iv). Therefore

$$\lim_k \|z_k - z_{k+1}\| = \lim_k \|w_k - w_{k+1}\|$$

since $\lim_k \|B^k A^t (y^o - y^\varepsilon)\| = 0$ by assumption (9.30).

This proves part iii).

$\square$

Theorem 9.11

Let the tolerance level satisfy the following conditions:

$$\lim_{\varepsilon,\beta \to o} \tau(\varepsilon,\beta) = 0 \, , \quad \lim_{\varepsilon,\beta \to o} \frac{\tau(\varepsilon,\beta)}{\varepsilon + 2\beta} > 1 .$$

Then if $z_{\varepsilon,\beta} := z_{h(\tau(\varepsilon,\beta))}$ is constructed according to the stopping rule (9.31) we have

i) $\quad h(\tau(\varepsilon,\beta)) \leq \dfrac{\|z_o - x^o\|^2}{(\tau(\varepsilon,\beta) - \varepsilon - 2\beta)(\tau(\varepsilon,\beta) - \varepsilon)} \quad .$

ii) $\quad \lim_{\varepsilon,\beta \to o} z_{\varepsilon,\beta} = x^o .$

Proof: Let $\Delta_k := w_k - x^o$, $\tilde{\Delta}_k := z_k - x^o$, $k \in \mathbb{N}$. Define $1 := 1(\varepsilon,\beta)$ by the following rule:

$$\|w_1 - w_{1+1}\| \leq \tau(\varepsilon,\beta) - \varepsilon, \quad \|w_k - w_{k+1}\| > \tau(\varepsilon,\beta) - \varepsilon$$

$$\text{for all } k < 1.$$

Since for all $k \in \mathbb{N}$

$$\|z_k - z_{k+1}\| \leq \|w_k - w_{k+1}\| + \|B^k A^t (y^o - y^\varepsilon)\| \leq \|w_k - w_{k+1}\| + \varepsilon$$

we have $h(\tau(\varepsilon,\beta)) \leq 1$. From

$$\|w_k - w_{k+1}\| > \tau(\varepsilon,\beta) - \varepsilon \text{ for all } k < 1$$

we obtain by i) in Lemma 9.10

$$l(\tau(\varepsilon,\beta) - \varepsilon - \beta)^2 \leq \sum_{j=0}^{l-1} \|w_j - w_{j+1} + u_j\|^2 \leq \|z_o - x^o\|^2 + l\beta^2$$

and therefore

$$l(\tau(\varepsilon,\beta) - \varepsilon - 2\beta)(\tau(\varepsilon,\beta) - \varepsilon) \leq \|z_o - x^o\|^2 ,$$

$$h(\tau(\varepsilon,\beta)) \leq l \leq \frac{\|z_o - x^o\|^2}{(\tau(\varepsilon,\beta) - \varepsilon - 2\beta)(\tau(\varepsilon,\beta) - \varepsilon)} .$$

This proves i).

Since

$$\tilde{\Delta}_{k+1} = B^{k+1}\tilde{\Delta}_o + \sum_{j=0}^{k} B^j \{A^t(y^\varepsilon - y^o) + u_{k-j}\}$$

we have

$$(*) \qquad \|\tilde{\Delta}_{k+1}\| \leq \|B^{k+1}\tilde{\Delta}_o\| + (k+1)(\varepsilon+\beta) , \quad k \in \mathbb{N}.$$

Now, let $(\varepsilon_n)_{n\in\mathbb{N}}$, $(\beta_n)_{n\in\mathbb{N}}$ are sequences with

$$\lim_n \varepsilon_n = 0 , \quad \lim_n \beta_n = 0.$$

Let $\tau_n := \tau(\varepsilon_n,\beta_n)$, $l_n := h(\tau_n)$ and $(z_k^n)_{k\in\mathbb{N}}$, $(w_k^n)_{k\in\mathbb{N}}$ constructed according to (9.29) and (9.32) respectively. We distinguish two cases:

a) $\underline{\lim}_n l_n \leq \bar{l} < \infty.$

Without loss of generality: $l_n \leq \bar{l}$ for all $n \in \mathbb{N}$.
Assume, by way of contradiction, there exists $\mu > o$ and $n_o \in \mathbb{N}$ such that

$$\|z_{l_n}^n - x^o\| \geq \mu > o \quad \text{for all } n \geq n_o.$$

Choose $\bar{k} \in \mathbb{N}$, $\bar{k} > 1$, and $N \geq n_o$ with

$$\tau_n \leq \frac{\mu}{2} , \quad \|B^{\bar{k}+1}\tilde{\Delta}_o\| < \frac{1}{4}\mu , \quad (\bar{k}+1)(\varepsilon_n+\beta_n) \leq \frac{\mu}{4}$$

for all $n \geq N$. Then by $(*)$

$$\|z_{\bar{k}+1}^n - x^o\| < \frac{\mu}{2} \quad \text{for all } n \geq N.$$

Since

$$\| z^n_{\overline{k}+1} - z^n_{1_n} \| > \frac{\mu}{2} \quad \text{for all } n \geq N$$

there must be for all $n \geq N$ a number j_n with

$$\| z^n_{j_n+1} - z^n_{j_n} \| \geq \frac{\mu}{2} \cdot \frac{1}{\overline{k} - 1_n} \geq \frac{\mu}{2\overline{k}} \ , \ j_n \geq 1_n.$$

This is a contradiction to the definition of 1_n (notice

$\tau_n < \frac{\mu}{2\overline{k}}$ for n large enough).

b) $\lim\limits_{n} 1_n = \infty.$

Since $\lim\limits_{n} \| B^{1_n} \tilde{\Delta}_o \| = 0$ it is sufficient to show that

$\lim\limits_{n} 1_n (\varepsilon_n + \beta_n) = 0$ because of (*). By assumption, there

exists $n \in \mathbb{N}$ with

$$\tau_n > \delta_n + 2\beta_n \quad \text{for all} \quad n \geq N.$$

Therefore by i)

$$(\#) \qquad 1_n \leq \frac{\| z_o - x^o \|^2}{(\tau_n - \varepsilon_n - 2\beta_n)(\tau_n - \varepsilon_n)} =: s_n, \ n \geq N.$$

We have

$$z^n_k - z^n_{k+1} = B^k (I-B)(z_o - x^o) - B^k A^t (y^\varepsilon - y^o)$$

$$+ \sum_{j=0}^{k-1} B^j (I-B) u_{k-1-j} - u_k$$

and therefore

$$\tau_n < \| z^n_{1_n-1} - z^n_{1_n} \| \leq \| B^{\frac{1}{2}(1_n-1)} B^{\frac{1}{2}(1_n-1)} (I-B)(z_o - x^o) \|$$

$$+ \varepsilon_n + \sum_{j=1}^{1_n-2} \| B^j (I-B) u_{1_n-2-j} \| + \beta_n$$

$$\leq \frac{2}{1_n} \| B^{\frac{1}{2}(1_n-1)} (z_o - x^o) \| + \varepsilon_n + \beta_n + \beta_n \sum_{j=1}^{1_n-2} \frac{1}{j}$$

$$\leq \frac{2}{l_n} \| B^{\frac{1}{2}(l_n-1)} (z_o - x^o) \| + \varepsilon_n + (2 + \ln(l_n))\, \beta.$$

With ($\neq$) we obtain

$$(\varepsilon_n + \beta_n)\,(l_n \tau_n - l_n(\varepsilon_n + (2 + \ln(s_n))\,\beta_n))$$

$$\leq 2 \| B^{\frac{1}{2}(l_n-1)} (z_o - x^o) \| \,(\varepsilon_n + \beta_n)$$

which gives

$$(\varepsilon_n + \beta_n) l_n \leq \frac{2 \| B^{\frac{1}{2}(l_n-1)} (z_o - x^o) \| \,(\varepsilon_n + \beta_n)}{\tau_n - \varepsilon_n - (2 + \ln(s_n))\,\beta_n} \,.$$

Since $\lim\limits_{n} \dfrac{\tau_n}{\varepsilon_n + \beta_n} > 1$ and $\lim\limits_{n} \| B^{\frac{1}{2}(l_n-1)} (z_o - x^o) \| = 0$

it follows $\lim\limits_{n} (\varepsilon_n + \beta_n) l_n = 0.$

$\square$

9.5 The ART-algorithm

Suppose that the system of equations is given by

$$(9.33) \qquad a_i^t x = y_i , \quad 1 \leq i \leq m ,$$

where $a_1, \ldots, a_m \in \mathbb{R}^n$ and $y_1, \ldots, y_m \in \mathbb{R}$. In practice, the pair (a_i, y_i) may be considered as a representation of one experiment:

y_i is the result of the experiment,

a_i describes the "geometry" of the experiment,

x is the unknown quantity to be estimated.

Let $J(x) := \sum\limits_{i=1}^{m} |a_i^t x - y_i|^2 $, $x \in \mathbb{R}^n$. The procedure that we use to find x^* such that

$$(9.34) \qquad J(x^*) = \min\{J(x) \mid x \in \mathbb{R}^n\}$$

is iterative and of adaptive type: In any step, an estimation of x at the next iteration step is constructed from that at the preceding step and from a new "observation" given by a pair (a_i, y_i); adaption is done by moving a certain (small) step in the direction opposite to the current gradient of the objective function J. This procedure leads to the following form of a recursive algorithm:

Choose $x_o \in \mathrm{IR}^n$, continue with

$$(9.35) \qquad x_{k+1} := x_k + \lambda_k (y_k - a_k^t x_k) a_k, \quad k \in \mathrm{IN}.$$

Here $(\lambda_k)_{k \in \mathrm{IN}}$ is a sequence of relaxation parameters and the data (a_i, y_i) are used in a cyclic order:

$$a_k = a_j, \quad y_k = y_j, \quad \text{if } j - 1 = k \bmod m.$$

In the literature an algorithm of this type is called an ART-algorithm (algebraic reconstruction technique). This type of algorithm is frequently used in reconstruction of images (see Examples (E3) and (E4) in Chapter 1).

To simplify in the sequel the computations we assume that the vectors a_j are normalized:

$$a_j^t a_j = 1 \quad , \quad 1 \leq j \leq m.$$

Moreover, we consider cyclic relaxation only, that is λ_k is constant during a cycle. Then we obtain from (9.35) the following iteration scheme:

Choose $x_o \in \mathrm{IR}^n$, continue with

$$(9.36) \qquad x_{km+j} := x_{km+j-1} + \lambda_k (y_j - a_j^t x_{km+j-1}) a_j \ ,$$

$$1 \leq j \leq m, \ k \in \mathrm{IN}.$$

Let $\tilde{x}$ be any solution of the problem (9.34) and let α be the corresponding residuum, that is

$$\alpha = J(\tilde{x}) = \min \{J(x) \mid x \in \mathrm{IR}^n\}.$$

Then from (9.36) we obtain

$$\|x_{km+j} - \tilde{x}\|^2 = \|x_{km+j-1} - \tilde{x}\|^2 + \lambda_k^2 |y_j - a_j^t x_{km+j-1}|^2$$

$$(9.37) \qquad + 2\lambda_k (y_j - a_j^t x_{km+j-1})(a_j^t x_{km+j-1} - y_j + y_j - a_j^t \tilde{x})$$

$$\leq \|x_{km+j-1} - \tilde{x}\|^2 - \sigma_k |y_j - a_j^t x_{km+j-1}|^2 + \lambda_k |y_j - a_j^t \tilde{x}|^2$$

with

$$\sigma_k := \begin{cases} \lambda_k (2 - \lambda_k) & , \ \text{if} \ \ \alpha = 0 \\ \lambda_k (1 - \lambda_k) & , \ \text{if} \ \ \alpha > 0 \end{cases} .$$

(In the case $\alpha > 0$ we have used the well known inequality $ab \leq \frac{1}{2}(a^2 + b^2)$). This implies

$$(9.38) \qquad \|x_{(k+1)m} - \tilde{x}\|^2 \leq \|x_{km} - \tilde{x}\|^2 - \sigma_k c_k + \lambda_k \alpha$$

where

$$c_k := \sum_{j=1}^{m} |y_j - a_j^t x_{km+j-1}|^2 \ , \ k \in \mathbb{N}.$$

An immediate consequence of the estimate (9.38) is the following theorem which gives some insight how the iteration method (9.36) works in the average.

Theorem 9.12

Let the sequence $(x_l)_{l \in \mathbb{N}}$ be determined by the iteration (9.36) with $\lambda_k = \lambda, k \in \mathbb{N}; \lambda \in (0,1)$. Then

$$\frac{1}{k} \left(\sum_{l=1}^{m} \left\{ \sum_{j=1}^{m} |y_j - a_j^t x_{km+j-1}|^2 \right\} \right) \leq \frac{1}{k} \cdot \frac{1}{\lambda(1-\lambda)} \|x_o - A^+ y\| + \frac{\alpha}{1-\lambda} ,$$

$$k \in \mathbb{N}.$$

<u>Proof:</u> We choose $\tilde{x} := A^+ y$ in (9.38) and obtain

$$\|x_{(k+1)m} - A^+ y\|^2 \leq \|x_o - A^+ y\|^2 - \lambda(1-\lambda) \sum_{l=1}^{k} c_l + \lambda k \alpha$$

and therefore

$$\sum_{l=1}^{k} c_l \leq \frac{1}{\lambda(1-\lambda)} \|x^o - A^+ y\|^2 + \frac{k}{1-\lambda} - \frac{1}{\lambda(1-\lambda)} \|x_{(k+1)m} - A^+ y\|^2$$

which implies the desired inequality.

$$\square$$

The main result in this section is

Theorem 9.13

Let the sequence $(x_1)_{1\in\mathbb{N}}$ be determined by the iteration (9.36) and suppose that $x_o \in R(A^t)$. Then we have

a) If the system is consistent ($\alpha = 0$) and if the relaxation sequence $(\lambda_k)_{k\in\mathbb{N}}$ satisfies

$$0 \le \lambda_k \le 2, \ k \in \mathbb{N}, \ \sum_{k=o}^{\infty} \lambda_k(2 - \lambda_k) = \infty$$

then $A^+y = \lim_k x_{km}$.

b) If the system is inconsistent ($\alpha > 0$) and if the relaxation sequence $(\lambda_k)_{k\in\mathbb{N}}$ satisfies

$$0 \le \lambda_k, \ k \in \mathbb{N}, \ \sum_{k=o}^{\infty} \lambda_k = \infty, \ \sum_{k=o}^{\infty} \lambda_k^2 < \infty$$

then $A^+y = \lim_k x_{km}$.

Proof: From the assumption $x_o \in R(A^t)$ and the identity $N(A)^{\perp} = R(A^t)$ we obtain

$$(1) \qquad x_1 \in N(A)^{\perp}, \ v_1 := x_1 - A^+y \in N(A)^{\perp}, \ 1 \in \mathbb{N}.$$

Setting $b_k := \|x_{km} - A^+y\|^2$, $c_k := \sum_{j=1}^{m} |y_j - a_j^t x_{km+j-1}|^2$, $k \in \mathbb{N}$,

(9.38) may be rewritten as follows

$$(2) \qquad \lambda_k(\mu_k c_k - \alpha) \le b_k - b_{k+1} \ , \ k \in \mathbb{N}$$

where

$$\mu_k := \begin{cases} 2 - \lambda_k \ , & \text{if } \alpha = 0 \\ 1 - \lambda_k \ , & \text{if } \alpha > 0 \end{cases} .$$

Let us prove part a).

From the inequality (2) we obtain that the sequence $(b_k)_{k\in\mathbb{N}}$ is monotone decreasing and therefore convergent. Now assume, by way of contradiction, that $b := \lim_k b_k > 0$. Then from (2)

$$0 \le \lambda_k(2-\lambda_k) \sum_{j=1}^{m} |a_j^t v_{km+j-1}|^2 \|v_{km}\|^{-2} \le 1 - \|v_{(k+1)m}\|^2 \|v_{km}\|^{-2} \le 1$$

for all $k \in \mathbb{N}$. Since $\sum\limits_{k=0}^{\infty} \lambda_k (2 - \lambda_k) = \infty$ and

$$\|v_{km+j-1}\|^2 \|v_{km}\|^{-2} \leq 1, \quad 1 \leq j \leq m, \; k \in \mathbb{N},$$

(see (9.37)) there must exist a subsequence $(k_p)_{p \in \mathbb{N}}$ and vectors $w_j \in N(A)^{\perp}$, $1 \leq j \leq m$, such that

$$(3) \qquad \lim_p v_{k_p m+j-1} \|v_{k_p m}\|^{-1} = w_{j-1}, \quad |w_{j-1}\| = 1, \; 1 \leq j \leq m,$$

$$(4) \qquad \lim_p \sum_{j=1}^{m} |a_j^t v_{k_p m+j-1}|^2 \|v_{k_p m}\|^{-2} = \sum_{j=1}^{m} |a_j^t w_{j-1}|^2 = 0.$$

Since the sequence $(\lambda_k)_{k \in \mathbb{N}}$ is bounded we obtain from (3) and (9.37) that $w_o = \ldots = w_{m-1}$. By (3) and (4) it follows

$$w_o \in N(A)^{\perp} \cap N(A), \quad \|w_o\| = 1$$

which is a contradiction.

Let us now prove part b).

The inequality (2) suggests to consider the following two cases separately:

$$\varliminf_k (c_k - \lambda_k c_k) > \alpha, \quad \varliminf_k (c_k - \lambda_k c_k) \leq \alpha.$$

Suppose that $\varliminf\limits_k (c_k - \lambda_k c_k) > \alpha$. Then we may assume (without loss of generality) that

$$(1 - \lambda_k) \, c_k \geq \alpha + \delta$$

for all $k \in \mathbb{N}$ for some $\delta > 0$. But the inequality

$$\delta \sum_{k=0}^{1} \lambda_k \leq \sum_{k=0}^{1} \lambda_k \{ (1-\lambda_k) c_k - \alpha \} \leq b_o, \quad 1 \in \mathbb{N}$$

which follows from (9.38) shows that this case cannot occur since $\sum\limits_{k=0}^{\infty} \lambda_k = \infty$.

Now consider the case $\varliminf\limits_k (c_k - \lambda_k c_k) \leq \alpha$. There exists a subsequence $(k_p)_{p \in \mathbb{N}}$ such that

$$(1 - \lambda_{k_p})c_{k_p} \leq \alpha, \quad p \in \mathbb{N}, \quad c_k - \lambda_k c_k \geq \alpha \text{ if } k \neq k_p.$$

Now assume, by way of contradiction, that $(x_{k_p m})_{p \in \mathbb{N}}$ is unbounded. Then there exists a subsequence of $(k_p)_{p \in \mathbb{N}}$, say $(k'_p)_{p \in \mathbb{N}}$, such that

$$0 \leq \tfrac{1}{2} c_{k'_p} \| x_{k'_p} \|^{-2} = \frac{1}{2} \sum_{j=1}^{m} | y_j - a_j^t x_{k'_p m+j-1} |^2 \| x_{k'_p m} \|^{-2} \leq \alpha \| x_{k'_p m} \|^{-2}$$

for all k'_p and such that

$$\lim_p x_{k'_p m} \| x_{k'_p m} \|^{-1} = w, \quad \| w \| = 1.$$

From the iteration scheme (9.36) it follows

$$\lim_p x_{k'_p m+j-1} \, | x_{k'_p m} |^{-1} = w, \quad 1 \leq j \leq m.$$

Hence

$$\sum_{j=1}^{m} | a_j^t w |^2 = \lim_p \sum_{j=1}^{m} | y_j - a_j^t x_{k'_p m+j-1} |^2 \, \| x_{k'_p m} \|^{-2} = 0,$$

$$w \in N(A) \cap N(A)^{\perp}, \quad |w| = 1,$$

which is a contradiction. Therefore $(x_{k_p})_{p \in \mathbb{N}}$ is bounded. If x is a cluster point of $(x_{k_p m})_{p \in \mathbb{N}}$ then we know $x \in N(A)^{\perp}$ and it is easy to see by similar arguments as above that x is a least squares solution ($J(x) = \alpha$); therefore $x = A^+ y$. This implies

$$\lim_p x_{k_p m} = A^+ y.$$

Since

$$b_{k+1} \leq b_k - \lambda_k ((1 - \lambda_k)c_k - \alpha) \leq b_k, \quad k \neq k_p,$$

we have $A^+ y = \lim_k x_{km}$.

$$\square$$

Corollary 9.14

Under the assumptions of Theorem 9.13 we have

$$A^+ y = \lim_k x_{km+j-1} \ , \quad 1 \leq j \leq m.$$

Proof: This is an immediate consequence of the results in Theorem 9.13 and the iteration scheme (9.36).

$\square$

The attractivity of ART-type algorithms comes from the following facts:

- The iteration step is simple to realize.

- No extra effort is necessary to add data (a_i, y_i).

The disadvantage of this type of iterative algorithms is that the convergence is in general slow.

Illustration 9.15

Let a_i, $1 \leq i \leq 6$, are the rows of the matrix

$$A := \begin{pmatrix} 1 & 3 & 2 & -1 \\ 1 & 2 & -1 & -2 \\ 1 & -1 & 2 & 3 \\ 2 & 1 & 1 & 1 \\ 5 & 5 & 4 & 1 \\ 4 & -1 & 5 & 7 \end{pmatrix}$$

and let $y^t = (5\ 0\ 5\ 5\ 15\ 15)$.

The rank of the matrix is three and the solution manifold of the equation $Ax = y$ is given by

$$A^+ y + \lambda z \ , \quad \lambda \in \mathbb{R},$$

where

$$(A^+ y)^t = \frac{5}{13}\ (3\ 2\ 3\ 2), \quad z^t = \frac{1}{13}\ (-5\ 3\ -5\ 3).$$

Results with the zero vector as initial guess x_o are shown in

Table 2. (Notice that the rows of A are normalized during the computation).

relaxation	cycle number	result			
$\lambda_k = 1$	30	1.15381	0.76924	1.15388	0.76923
	40	1.15384	0.76923	1.15385	0.76923
$\lambda_k = 0.5$	30	1.15283	0.76953	1.15498	0.76902
	60	1.15384	0.76923	1.15386	0.76923
$\lambda_k = \frac{1}{k}$	30	1.01019	0.80550	1.30464	0.73772
	100	1.04615	0.79595	1.26622	0.76923
A^+y		1.15384	0.76923	1.15384	0.76923

Table 2

Bibliographical comments

An excellent treatment of numerical aspects of least squares problems is given in GOLUB, VAN LOAN [41]. Our discussion of the factorization approach and the problem of rank decision is drawn from papers by SAUTTER [91] and DEUFLHARD, SAUTTER [26]. More on cross-validation can be found in GOLUB, HEATH, WHABA [40] and ELDEN [31]; the optimal smoothing of data using cross-validation is described in CRAVEN, WHABA [23] and UTRERAS [99]. The results in Section 9.4 are taken from EMELIN, KRASNOSELSKII [33]; see also VERETENNIKOV, KRASNOSELSKII [100]. The ART-algoritm, originally proposed by KACZMARZ [61], was rediscovered several times; our presentation follows BAUMEISTER [9]. A variant for nonlinear systems is described by MARTINEZ [68].

PART IV

SPECIFIC TOPICS

The final part of these notes is devoted to the discussion
of some specific inverse problems: inverse problems which are
governed by convolution equations, evolution problems backwards in time, parameter identification.

Chapter 10: Convolution equations

In this chapter we shall consider problems in which the
Fourier transform plays an essential role: convolution equa-
tions, reconstruction from projections (see examples (E2),
(E3) in Chapter 1).

10.1 The Fourier transform

In this section we mention the necessary information from
the theory of Fourier transforms.

Let $f : \mathbb{R}^n \longrightarrow \mathbb{C}$. We define functions $\hat{f}, \check{f} : \mathbb{R}^n \longrightarrow \mathbb{C}$
formally by

$$\hat{f}(w) := (2\pi)^{-\frac{n}{2}} \int_{\mathbb{R}^n} f(t) \, \exp((-iw,t))dt, \; w \in \mathbb{R}^n,$$

$$\check{f}(t) := (2\pi)^{-\frac{n}{2}} \int_{\mathbb{R}^n} f(w) \, \exp((+iw,t))dw, \; t \in \mathbb{R}^n;$$

here $(\cdot,\cdot)$ denotes the euclidean inner product in $\mathbb{R}^n$.

It is clear that the integrals above exist if $f \in L_1(\mathbb{R}^n)$.
Therefore the transformations " $\wedge$ " and " $\vee$ " define linear
mappings on $L_1(\mathbb{R}^n)$. Moreover, it is easy to establish the
following result:

Lemma of Riemann-Lebesgue

If $f \in L_1(\mathbb{R}^n)$ then $\hat{f} \in C_o(\mathbb{R}^n)$ where $C_o(\mathbb{R}^n)$ is the space
$\{h : \mathbb{R}^n \longrightarrow \mathbb{C} \mid h$ is uniformly continuous, $\lim_{|t| \to \infty} h(t) = 0\}$.

We shall also consider the transformations "$\wedge$" and "$\vee$" on the space $L_2(\mathbb{R}^n)$ (with inner product $(.,.)$ and norm $\| \; \|$). How this is possible becomes clear from

Plancherel's theorem

There exist uniquely determined bounded linear operators F, $F_- : L_2(\mathbb{R}^n) \longrightarrow L_2(\mathbb{R}^n)$ such that the following assertions are true:

(1) $F(f) = \hat{f}$, $F_-(f) = \check{f}$ for all $f \in L_2(\mathbb{R}^n) \cap L_1(\mathbb{R}^n)$.

(2) $R(F) = R(F_-) = L_2(\mathbb{R}^n)$.

(3) $FF_- = F_-F = I$.

(4) $(F(f), F(g)) = (f,g)$ for all $f,g \in L_2(\mathbb{R}^n)$.

(5) $\|F(f)\| = \|f\|$, $\|F_-(f)\| = f$ for all $f,g \in L_2(\mathbb{R}^n)$.

The identities (2) and (3) may be summarized by saying that F^{-1} exists and is given by F_-. From now we do not distinguish between the transforms F and F^{-1} and "$\wedge$" and "$\vee$" respectively. In the following "$\wedge$" is called the Fourier transform and "$\vee$" is called the inverse

The convolution $f * g$ of functions f,g on $\mathbb{R}^n$ is defined by

$$(f * g)(t) := \int_{\mathbb{R}^n} f(t-s)g(s)\,ds \quad , \; t \in \mathbb{R}^n .$$

Convolution theorem

Let $f \in L_1(\mathbb{R}^n)$ and $g \in L_2(\mathbb{R}^n)$. Then

$$f * g \in L_2(\mathbb{R}^n) \text{ and } \widehat{f * g} = (2\pi)^{\frac{n}{2}} \hat{f}\,\hat{g}$$

This convolution theorem in connection with the lemma of Riemann-Lebesgue shows very clear that a convolution equation

$$(10.1) \qquad \kappa * x = y \quad (\kappa \in L_1(\mathbb{R}^n))$$

considered as an equation of $L_2(\mathbb{R}^n)$ into $L_2(\mathbb{R}^n)$ is ill-posed due to the lack of stability:

- If a solution x of (10.1) exists then
$$x = \overset{\vee}{f} \quad \text{where} \quad f = \hat{y} \, \overset{\wedge}{\kappa}^{-1}.$$

- A small perturbation g in y whoose transform $\hat{g}$ does not decay faster than $\overset{\wedge}{\kappa}$ as $|\omega| \to \infty$ will result in a perturbation in $\hat{y} \, \overset{\wedge}{\kappa}^{-1}$ which will grow without bound.

Notice that the operator $A : L_2(\mathbb{R}^n) \longrightarrow L_2(\mathbb{R}^n)$ associated with the kernel κ is not compact. But the Fourier transform plays the role of the singular value decomposition for compact operators.

Let us consider the Fourier transform of special functions which are of some interest in the following. In these examples it is useful to consider the transformation "$\wedge$" as a transformation of the "state space with time t" into the "frequency space with frequency ω".

Example 10.1

1) The perfect lowpass filter (see Example 4.4). Let
$$h_1(t) := \frac{\Omega}{\sqrt{2\pi}} \, \frac{\sin(\frac{1}{2}\Omega t)}{\frac{1}{2}\Omega t} \, , \quad t \in \mathbb{R}.$$

Then
$$\overset{\wedge}{h_1}(\omega) = \begin{cases} 1 & , \quad |\omega| \leq \Omega \\ 0 & , \quad |\omega| > \Omega \end{cases} .$$

2) The triangle window. Let
$$h_2(t) := \frac{1}{\sqrt{2\pi}} \left\{ \frac{\sin(\frac{1}{2}\Omega t)}{\frac{1}{2}\Omega t} \right\}^2 \, , \quad t \in \mathbb{R}.$$

Then

$$\hat{h}_2(\omega) = \begin{cases} \frac{1}{\Omega}(1 - \frac{|\omega|}{\Omega}) & , \ |\omega| \leq \Omega \\ 0 & \ |\omega| > \Omega. \end{cases}$$

This can be easily verified by using the convolution theorem since

$$h_2 = \sqrt{2\pi} \ \frac{1}{\Omega^2} \ h_1^2 \ .$$

3) The Lorentz-function. Let

$$h(t) := \frac{a}{t^2+a^2} \ , \quad t \in \mathbb{R}.$$

Then

$$\hat{h}(\omega) = \sqrt{\frac{\pi}{2}} \ e^{-a\omega} \ , \quad \omega \in \mathbb{R}. \qquad *$$

Since a convolution equation of the form (10.1) is ill-posed we have to regularize the equation in order to solve it in a stable way. If we consider the equation (10.1) in the frequency domain then regularization may be described by a filter function (window function) h in the following way:

$$x_h = \overset{\vee}{f_h} \ , \quad \overset{\wedge}{f_h} = \hat{h} \ \hat{y} \overset{\wedge}{\kappa}{}^{-1}.$$

If we consider the filter function $h = h_1$ (see Ex.10.1) $\hat{x}_h$ has the truncated frequency spectrum

$$\hat{x}_h(\omega) = \begin{cases} \hat{x}(\omega) & , \ |\omega| \leq \Omega \\ 0 & , \ |\omega| > \Omega \end{cases} \ .$$

The choice of the filter function h has to ensure that $f_h = \hat{h} \ \hat{y} \overset{\wedge}{\kappa}{}^{-1}$ doesn't blow up for $|\omega| \to \infty$. In general the filter function depends on a parameter which has to be chosen properly. In the next section we shall consider a filter function which is related to Tikhonov's regularization.

10.2 Regularization of convolution equations: Asymptotic

estimates

We consider a one-dimensional copy of the equation (10.1)

$$(10.2) \quad \kappa * x = y$$

where the kernel κ is a given real valued function in $L_1(\mathbb{R})$.

In the usual setting, let us assume:

There are given real valued functions

$$(10.3) \quad x^o, y^o, y^\varepsilon \in L_2(\mathbb{R}) \quad \text{with}$$

$$\kappa * x^o = y^o, \; \|y^o - y^\varepsilon\| \le \varepsilon.$$

As we know from the last section, the convolution equation
(10.2) is equivalent to the "algebraic" equation

$$\sqrt{2\pi} \; \hat{\kappa} \hat{x} = \hat{y}.$$

Since $\lim\limits_{|\omega| \to \infty} |\hat{\kappa}(\omega)| = 0$ it is not reasonable to define the reconstruction x^ε of x^o as

$$x^\varepsilon := \frac{1}{\sqrt{2\pi}} \; \overset{\vee}{f} \quad \text{where } f = \hat{y}^\varepsilon \, \hat{\kappa}^{-1}.$$

Here we consider the following family $(x^{\varepsilon,\alpha})_{\alpha>o}$ of regularized
solutions of the reconstruction problem:

$$x^{\varepsilon,\alpha} := \frac{1}{\sqrt{2\pi}} \; \overset{\vee}{f}^{\varepsilon,\alpha}$$

$$\text{where } f^{\varepsilon,\alpha}(\omega) := \frac{|\hat{\kappa}(\omega)|^2}{|\hat{\kappa}(\omega)|^2 + \alpha(1+\omega^2)} \cdot \frac{\hat{y}^\varepsilon(\omega)}{\hat{\kappa}(\omega)} \;, \omega \in \mathbb{R}.$$

This regularized solution $x^{\varepsilon,\alpha}$ may equivalentely be defined as
the minimizer of the Tikhonov-functional

$$x \longmapsto \|\kappa * x - y^\varepsilon\|^2 + \alpha |x|_1^2$$

where $\| \; \|_1$ is the norm in the Sobolev space

$$H_1(\mathbb{R}) := \{x \in L_2(\mathbb{R}) \mid x \text{ absolutely continuous, } x' \in L_2(\mathbb{R})\}.$$

In the following we shall give a bound for the reconstruction error $x^{\varepsilon,\alpha} - x^o$. By Plancherel's theorem it is sufficient to estimate $f^{\varepsilon,\alpha} - f^o$ where $f^o = \hat{x}^o$. Let

$$z(\omega;\alpha) := \frac{|\hat{\kappa}(\omega)|^2}{|\hat{\kappa}(\omega)|^2 + \alpha(1+\omega^2)} \quad , \quad v(\omega;\alpha) := \frac{1+\omega^2}{|\hat{\kappa}(\omega)|^2 + \alpha(1+\omega^2)} \quad ,$$

and

$$f^{o,\alpha}(\omega) := z(\omega;\alpha)\frac{\hat{y}^o(\omega)}{\hat{\kappa}(\omega)} = z(\omega;\alpha)\hat{x}^o(\omega) \quad , \quad \omega \in \mathbb{R} .$$

We estimate $f^{\varepsilon,\alpha} - f^o$ by estimating each term on the right-hand side of

$$\|f^{\varepsilon,\alpha} - f^o\| \le \|f^{\varepsilon,\alpha} - f^{o,\alpha}\| + \|f^{o,\alpha} - f^o\| .$$

Notice that the functions

$$\omega \longmapsto |\hat{x}^o(\omega)| \quad , \quad \omega \longmapsto |\hat{\kappa}(\omega)|$$

are even function since x^o and κ are real valued.

<u>Lemma 10.2</u>

We have

$$\|f^{\varepsilon,\alpha} - f^{o,\alpha}\| \le \frac{\varepsilon}{\sqrt{\alpha}} .$$

<u>Proof:</u>

$$\|f^{\varepsilon,\alpha} - f^{o,\alpha}\|^2 = \int_{-\infty}^{\infty} z(\omega;\alpha)^2 \frac{1}{|\hat{\kappa}(\omega)|^2} |\hat{y}^\varepsilon(\omega) - \hat{y}^o(\omega)|^2 d\omega$$

$$\le \frac{1}{\alpha} \int_{-\infty}^{\infty} |\hat{y}^\varepsilon(\omega) - \hat{y}^o(\omega)|^2 d\omega$$

$$= \frac{1}{\alpha} \int_{-\infty}^{\infty} |y^\varepsilon(t) - y^o(t)|^2 dt$$

$$\le \frac{\varepsilon^2}{\alpha} \quad . \qquad \qquad \square$$

As we already know the bound for $f^{o,\alpha} - f^o$ depends on an a-priori information about the exact solution x^o. Here we assume:

(10.3)   There exists a number $q > \frac{1}{2}$ and a constant $c_1 > 0$ such that

$$|\hat{x}^o(\omega)| \le c_1 (1 + \omega^2)^{-\frac{q}{2}} , \quad \omega \in \mathbb{R}.$$

The second ingredient for deriving an estimate for $f^{o,\alpha} - f^o$ consists in an information on the "order of ill-posedness" of the equation. As it is suggested by the algebraic equation

$$\sqrt{2\pi}\ \hat{\kappa}\ \hat{x}^o = \hat{y}^o$$

this order of ill-posedness depends essentially on the behaviour of $\hat{\kappa}$ in the neighbourhood of its zeros and at infinity. We shall consider two types of kernels. Concerning the first type we assume:

(1) $|\kappa(o)| > 0$.

(2) $\hat{\kappa}$ has at most a countable number of zeros ω_j, $j \in J$, with no accumulation point in $(0,\infty)$ if J is infinite. If $N(M)$ denotes the number of zeros of $\hat{\kappa}$ in $(0,M)$ then

(10.4) $$n(M) \le c_2 M^\gamma$$

with $\gamma \ge 0$, $c_2 \ge 0$.

(3)   There exist nonoverlapping open intervalls $D_j \subset (0,\infty)$ with $\omega_j \in D_j$, $j \in J$, a number $p > \frac{1}{4}$ and a constant $c_3 > 0$ such that

$$|\hat{\kappa}(\omega)|^2 \ge c_3 |\omega - \omega_j|^{2p} \text{ for } \omega \in D_j, \ j \in J.$$

(4)   There exist a number $a > 0$ and a constant $c_4 > 0$ such that

$$|\kappa(\omega)|^2 \ge c_4 (1 + \omega^2)^{-2a} \text{ if } \omega \notin \bigcup_{j \in J} D_j \ .$$

Let $M \geq 1$, $N := N(M)$ and $\omega_0 < \ldots < \omega_N$ be the zeros of $\hat{\kappa}$ in $(0,M)$. Then

$$\frac{1}{2}\|f^{0,\alpha} - f^0\|^2 = \frac{1}{2}\,\alpha^2 \int_{-\infty}^{\infty} \nu(\omega;\alpha)^2 |\hat{x}^0(\omega)|^2 d\omega$$

$$(10.5) \quad = \alpha^2 \int_{0}^{M} \nu(\omega;\alpha)^2 |\hat{x}^0(\omega)|^2 d\omega + \alpha^2 \int_{M}^{\infty} \nu(\omega;\alpha)^2 |\hat{x}^0(\omega)|^2 d\omega$$

$$=: J_1(M) + J_2(M).$$

Lemma 10.3

Suppose that the assumptions (10.3) and (10.4) are satisfied. Then

$$J_2(M) \leq c_5 M^{-2q+1}.$$

Proof:

$$J_2(M) \leq \alpha^2 \int_{M}^{\infty} \frac{(1+\omega^2)^2}{\alpha^2(1+\omega^2)^2} |\hat{x}^0(\omega)|^2 d\omega$$

$$\leq \int_{M}^{\infty} c_1^2 (1+\omega^2)^{-q} d\omega \leq c_1^2 \int_{M}^{\infty} \omega^{-2q} d\omega. \qquad \square$$

Let $D_j = (\omega_j - d_j,\ \omega_j + d_j)$, $1 \leq j \leq N$, and let $D_j' := (\omega_j + d_j, \omega_{j+1} - d_{j+1})$, $1 \leq j \leq N-1$. We split $J_1(M)$:

$$J_1(M) = \alpha^2 \int_{0}^{d_1} \nu(\omega;\alpha)^2 |\hat{x}^0(\omega)|^2 d\omega + \alpha^2 \sum_{j=1}^{N} \int_{D_j} \nu(\omega;\alpha)^2 |\hat{x}^0(\omega)|^2 d\omega$$

$$+ \sum_{j=1}^{N-1} \int_{D_j'} \nu(\omega;\alpha)^2 |\hat{x}^0(\omega)|^2 d\omega + \int_{\omega_N + d_N}^{M} \nu(\omega;\alpha)^2 |\hat{x}^0(\omega)|^2 d\omega$$

$$=: J^1 + J^2 + J^3 + J^4.$$

For J^1 we obtain

$$(10.6) \quad J^1 \leq c_6\,\alpha^2.$$

Now, we estimate a summand of J^2:

$$\alpha^2 \int_{D_j} v(\omega;\alpha)^2 |\hat{x}^0(\omega)|^2 d\omega$$

$$\leq \alpha^2 \int_{D_j} \frac{(1+\omega^2)^2 c_1^2 (1+\omega^2)^{-q}}{(c_3(\omega-\omega_j)^{2p} + \alpha)^2} d\omega$$

$$\leq c_1^2 \int_{D_j} \frac{(1+\omega^2)^2}{(c_3\alpha^{-1}(\omega-\omega_j)^{2p}+1)^2} d\omega$$

$$\leq c_1^2 (1+M^2)^2 \left(\frac{\alpha}{c_3}\right)^{1/2p} \int_{-\infty}^{\infty} \frac{d\mu}{(\mu^{2p}+1)^2}$$

$$\leq c_7 M^4 \alpha^{\frac{1}{2p}}.$$

Therefore

$$(10.7) \quad J^2 \leq c_7' M^{4+\gamma} \alpha^{\frac{1}{2p}} .$$

In a similar way we obtain bounds for J^3 and J^4:

$$(10.8) \quad J^3 \leq c_8 \alpha^2 M^{\gamma+5+8a} ,$$

$$(10.9) \quad J^4 \leq c_9 \alpha^2 M^{\gamma+5+8a} .$$

We summarize:

Lemma 10.4

If the assumptions (10.3) and (10.4) hold then

$$J_1(M) \leq c_{10}\{\alpha^2 M^{5+\gamma+8a} + \alpha^{1/2p} M^{4+\gamma}\} .$$

Theorem 10.5

If the assumptions (10.3) and (10.4) hold then

$$\|x^{\varepsilon,\alpha(\varepsilon)} - x^0\| \leq c\varepsilon^{\frac{2\tau}{2\tau+1}}$$

where $\tau = \dfrac{1}{p(2q-1)(4+2q+\gamma+8a)}$, $\alpha(\varepsilon) = \varepsilon^{\frac{2}{2\tau+1}}$.

Proof:

By Lemma 10.3 and Lemma 10.4 we obtain

$$\| f^{0,\alpha} - f^{0} \|^{2} \leq c_{11} \{ M^{-2q+1} + (\alpha^{2} + \alpha^{\frac{1}{2p}}) M^{5+\gamma+8a} \}.$$

Minimizing the right-hand side with respect to M and discarding terms of higher order in α we arrive at an estimate

$$\| f^{0,\alpha} - f^{0} \| \leq c_{12}\, \alpha^{\tau}$$

with

$$\tau = \frac{2q - 1}{4p(4+2q+\gamma+8a)}.$$

By Lemma 10.2 and Plancherel's theorem we obtain

$$\| x^{\varepsilon,\alpha} - x^{0} \| \leq c_{13} \{ \frac{\varepsilon}{\sqrt{\alpha}} + \alpha^{\tau} \}.$$

The parameter choice strategy

$$\alpha(\varepsilon) := \varepsilon^{\frac{2}{2\tau+1}}$$

leads to the result. $\square$

Remark 10.6

The result in Th.10.5 contains an a-priori strategy for the parameter choice problem. It shows also that the estimate for the rate of convergence decreases if the behaviour of $\hat{\kappa}^{-1}$ in the neighbourhood of the zeros of $\hat{\kappa}$ becomes more singular (parameter p!). *

Now we come to the second type of kernels which we want to consider. We change the assumption (10.4):

There exist a number a > o and a constant

(10.10) $c_{o} > 0$ such that

$$| \hat{\kappa}(\omega) |^{2} \geq c_{o} \exp(-a\omega), \quad \omega \geq 0.$$

A kernel which satisfies the assumption (10.10) is given by

the Lorentz-function (see Example 10.1).

To obtain an estimate for the reconstruction error $x^{\varepsilon,\alpha} - x^o$ we have only to give a new estimation for the integral $J_1(M)$ (see 10.5).

Lemma 10.7

If the assumptions (10.3) and (10.10) hold then

$$J_1(M) \leq c_{14}\, \alpha^2\, e^{2aM} \quad .$$

Proof: This follows immediately from the assumptions (10.3) and (10.10). $\square$

Theorem 10.8

Let the assumptions (10.3) and (10.10) hold. Then

$$\| x^{\varepsilon,\alpha(\varepsilon)} - x^o \| \leq c\, (\ln \tfrac{1}{\varepsilon})^{-q+\frac{1}{2}}$$

where $\alpha(\varepsilon) = \varepsilon^2\, \{\ln \tfrac{1}{\varepsilon^2}\}^{-q+\frac{1}{2}}$.

Proof: By Lemma 10.3 and Lemma 10.7 we have

$$(*) \qquad \| f^{o,\alpha} - f^o \|^2 \leq c_{15}\, \{M^{-2q+1} + \alpha^2 e^{2aM}\}.$$

With the choice

$$M = \frac{1}{a} \ln \frac{1}{\alpha} - \frac{2q-1}{2a} \ln(\tfrac{1}{a} \ln \tfrac{1}{\alpha})$$

as an approximation for the minimizer of the right-hand side in (*) with respect to M we arrive at

$$\| f^{o,\alpha} - f^o \| \leq c_{16}\, \{\ln \tfrac{1}{\alpha}\}^{-q+\frac{1}{2}}$$

and

$$\| x^{\varepsilon,\alpha} - x^o \| \leq c_{17}(\frac{\varepsilon}{\sqrt{\alpha}} + \{\ln \tfrac{1}{\varepsilon}\}^{-q+\frac{1}{2}}) \quad .$$

The choice

$$\alpha(\varepsilon) := \varepsilon^2 \{\ln \frac{1}{\varepsilon}\}^{2q-1}$$

leads to

$$\|x^{\varepsilon,\alpha} - x^o\| \le c_{18} (\ln \frac{1}{\varepsilon})^{-q + \frac{1}{2}} \quad . \qquad \square$$

10.3 On the discretization of convolution equations

Let us consider again a convolution equation $\kappa * x = y$ with $\kappa \in L_1(\mathbb{R})$ and let us assume:

There are given real valued functions
(10.11) $x^o, y^o, y^\varepsilon \in L_2(\mathbb{R})$ with

$$\kappa * x^o = y^o \;, \quad \|y^o - y^\varepsilon\| \le \varepsilon.$$

We shall look for a regularized (filtered) solution of the reconstruction problem within the space of trigonometric polynomials with period 2M, M > 0.

Let M > 0 and define

$$L_{2,M} := \{f : \mathbb{R} \longrightarrow \mathbb{C} \,|\, f \text{ 2M-periodic, } f\big|_{[-M,M]} \in L_2(-M,M)\}.$$

Then $L_{2,M}$ is a Hilbert space with inner product[*]

$$(f,g) = \frac{1}{2M} (\int_{-M}^{M} f(t) \, \overline{g(t)} \, dt)^{1/2}$$

and orthonormal basis $(e_k)_{k \in \mathbb{Z}}$ where

$$e_k(t) := \exp(ikt\frac{\pi}{M}) \quad , \; t \in \mathbb{R}.$$

Therefore, the Fourier coefficients of a function $f \in L_{2,M}$ are given by

$$\hat{f}(k) := (f,e_k) = \frac{1}{2M} \int_{-M}^{M} f(t) \, \overline{e_k(t)} \, dt, \; k \in \mathbb{Z} \;,$$

---

[*] If z is a complex number, then $\overline{z}$ denotes the complex conjugate of z.

and f may be represented by the Fourier series

$$\sum_{k \in \mathbb{Z}} \hat{f}(k) e_k.$$

Next, we replace in the computation of the Fourier coefficients the integration by a summation: Let $T = \dfrac{2M}{N}$, $N = 2N_1 \in \mathbb{N}$, be the "sampling" interval and let $\hat{f}_N(k)$ be computed by

$$\hat{f}_N(k) = \frac{1}{2M} \cdot T \cdot \sum_{n=-N_1}^{N_1-1} f(t_n)\, e_k(t_n)$$

$$= \frac{1}{N} \sum_{n=-N_1}^{N_1-1} f(t_n)\, e_k(t_n) \ , \quad k \in \mathbb{Z} \ ,$$

where $t_n = n \cdot T$, $n \in \mathbb{Z}$. Clearly, the sequence $(\hat{f}_N(k))_{k \in \mathbb{Z}}$ is periodic. With these "Fourier coefficients" we may define a trigonometric polynomial f_N by

$$f_N(t) := \sum_{k=-N_1}^{N_1-1} \hat{f}_N(k)\, e_k(t), \quad t \in \mathbb{R}.$$

Let us come back to the convolution equation $\kappa * x = y$ and let us describe the discretization.

First, let κ_M, x_M, y_M be periodic continuations of $\kappa|_{[-M,M]}$, $x|_{[-M,M]}$, $y|_{[-M,M]}$ respectively. We replace the convolution $\kappa * x$ by $\kappa_M \overset{M}{*} x_M$ defined by

$$\kappa_M \overset{M}{*} x_M (t) := \int_{-M}^{M} \kappa_M(t - s)\, x(s)\, ds \ , \quad t \in [-M,M].$$

Then if p,q are trigonometric polynomials considered as approximations for κ_M and x_M respectively we see immediately that in the computation of $p \overset{M}{*} q$ there are also used values of p outside of $[-M,M]$. But outside of $[-M,M]$ p is a bad approximation of κ since p is periodic and $\kappa \in L_1(\mathbb{R})$. Therefore, $p \overset{M}{*} q$ may be a bad approximation of $\kappa * x$ in $[-M,M]$. We take into consideration this fact by constructing p as an approximation of $\kappa \chi_{[-\frac{M}{2},\frac{M}{2}]}$ instead of $\kappa\big|_{[-M,M]}$.[*]

This leads us to consider the following approximation for the quantities κ and y^ε ($N_2 := \frac{1}{2}N_1$)

$$\kappa_{M,N}(t) := \sum_{k=-N_1}^{N_1-1} \hat{\kappa}_{M,N}(k)\,e_k(t),$$

$$\hat{\kappa}_{M,N}(k) := \frac{1}{N}\sum_{n=-N_2}^{N_2-1} \kappa(t_n)\,e_k(t_n), \quad -N_1 \le k \le N_1 - 1,$$

$$y^\varepsilon_{M,N}(t) := \sum_{k=-N_1}^{N_1-1} \hat{y}^\varepsilon_{M,N}(k)\,e_k(t),$$

$$\hat{y}^\varepsilon_{M,N}(k) := \frac{1}{N}\sum_{n=-N_1}^{N_1-1} y^\varepsilon(t_n)\,e_k(t_n), \quad -N_1 \le k \le N_1 - 1.$$

Then the reconstruction problem consists in finding a trigonometric polynomial

$$x_{M,N} = \sum_{k=-N_1}^{N_1-1} \hat{x}_{M,N}\,e_k$$

such that

$$(10.12) \qquad \kappa_{M,N} \overset{M}{*} x_{M,N} = y^\varepsilon_{M,N}.$$

The equation (10.12) is equivalent with

$$(10.13) \qquad 2M\,\hat{\kappa}_{M,N}(k)\,\hat{x}_{M,N}(k) = \hat{y}^\varepsilon_{M,N}(k), \quad -N_1 \le k \le N_1 - 1.$$

Clearly, the solution of (10.13) may be a very unreasonable solution due to the ill-posedness of the continuous problem. Therefore, we use Tikhonov regularization:

$$(10.14) \qquad \text{Let } x^\alpha_{M,N} = \sum_{k=-N_1}^{N_1-1} \hat{x}^\alpha_{M,N}\,e_k \text{ be the minimizer of the}$$

$$\text{function} \qquad \|\kappa_{M,N} \overset{M}{*} x_{M,N} - y^\varepsilon_{M,N}\|^2 + \alpha\|x^{(q)}_{M,N}\|^2$$

______

[*] χ_S is the characteristic function of the set S.

where $\alpha \geq 0$ and $q \in \mathbb{N}$. A simple calculation shows that $x_{M,N}^{\alpha}$ is given by

$$(10.15) \qquad x_{M,N}^{\alpha} = \sum_{k=-N_1}^{N_1-1} \hat{x}_{M,N}^{\alpha}(k)\, e_k$$

where

$$\hat{x}_{M,N}^{\alpha}(k) = \frac{2M|\hat{\kappa}_{M,N}(k)|^2}{(2M)^2|\hat{\kappa}_{M,N}(k)|^2 + \alpha\mu(k)^{2q}} \cdot \frac{\hat{y}_{M,N}^{\varepsilon}(k)}{\hat{\kappa}_{M,N}(k)} \quad ,$$

$$\mu(k) = k\,\frac{\pi}{M} \ , \quad -N_1 \leq k \leq N_1 - 1.$$

Let $\omega(N) := \exp(-i\,\frac{2\pi}{N})$. Using periodic properties we can write down the discrete quantities as follows:

$$(10.16) \quad \hat{\kappa}_{M,N}(k) = \frac{1}{N}\sum_{n=o}^{N-1} \kappa_n \omega(N)^{nk} \ , \quad 0 \leq k \leq N - 1,$$

$$\text{where } \kappa_n = \begin{cases} \kappa(t_n), & 0 \leq n \leq N_2 - 1, \ 3N_2 \leq n \leq N - 1 \\ 0, & \text{else ;} \end{cases}$$

$$(10.17) \quad \hat{y}_{M,N}^{\varepsilon}(k) = \frac{1}{N}\sum_{n=o}^{N-1} y^{\varepsilon}(t_n)\,\omega(N)^{nk} \ , \quad 0 \leq k \leq N - 1,$$

$$\hat{x}_{M,N}^{\alpha}(k) = \frac{2M|\hat{\kappa}_{M,N}(k)|^2}{(2M)^2|\hat{\kappa}_{M,N}(k)|^2 + \alpha\tilde{\mu}(k)^{2q}} \cdot \frac{\hat{y}_{M,N}^{\varepsilon}(k)}{\hat{\kappa}_{M,N}(k)} \quad ,$$

$$\text{with } \tilde{\mu}(k) = \begin{cases} \mu(k) & , \ 0 \leq k \leq N_1 - 1 \\ \mu(N-k) & , \ N_1 \leq k \leq N - 1 \ ; \end{cases}$$

$$(10.18) \quad x_{M,N}^{\alpha}(t_{N-j}) = \frac{1}{N}\sum_{k=o}^{N-1} (N\hat{x}_{M,N}^{\alpha}(k))\,\omega(N)^{jk}, \quad 0 \leq j \leq N - 1.$$

We see that the computations in (10.16),(10.17) and (10.18) are based on the same formula, namely:

$$\text{Given } \omega(N) := \exp(-i\,\frac{2\pi}{N}) \text{ and } z_o,\ldots,z_{N-1}.$$

$$\text{Compute } Z_k := \sum_{n=o}^{N-1} z_n \omega(N)^{kn}, \ 0 \leq k \leq N - 1.$$

The transform $(z_0, \ldots, z_{N-1}) \longmapsto (Z_0, \ldots, Z_{N-1})$ is called the discrete Fourier transform (DFT). As it is easily seen, the computation of $Z_0, \ldots, Z_{N-1}$ requires a number of arithmetic operations which is proportional to N^2. The fast Fourier transform (FFT) is a method which reduces the number of arithmetic operations to $N \log_2 N$ by using the following observation: A DFT of order N can be evaluated from two DFT of order $\frac{N}{2}$. Let us give a short sketch of this method.

Let $N = 2^\tau$, $\tau \geq 2$. We have with $N_1 := \frac{1}{2}N$

$$Z_k = \sum_{n=o}^{N-1} z_n \omega(N)^{kn}$$

$$= \sum_{r=o}^{N_1-1} z_{2r} \omega(N)^{2rk} + \sum_{r=o}^{N_1-1} z_{2r+1} \omega(N)^{(2r+1)k}$$

$$= \sum_{r=o}^{N_1-1} z_{2r} \omega(N_1)^{rk} + \omega(N)^k \sum_{r=o}^{N_1-1} z_{2r+1} \omega(N_1)^{rk}$$

and therefore

$$(10.19) \quad Z_k = U_k + \omega(N)^k V_k, \quad Z_{k+N_1} = U_k - \omega(N)^k V_k, \quad 0 \leq k \leq N_1 - 1$$

where

$$U_k = \sum_{r=o}^{N_1-1} z_{2r} \omega(N_1)^{rk}, \quad V_k = \sum_{r=o}^{N_1-1} z_{2r+1} \omega(N_1)^{rk},$$

$$0 \leq k \leq N_1 - 1.$$

Obviously, $U_0, \ldots, U_{N_1-1}$ and $V_0, \ldots, V_{N_1-1}$ are discrete Fourier transforms of $z_0, z_2, \ldots, z_{2(N_1-1)}$ and $z_1, \ldots, z_{N-1}$ respectively. If we apply the same procedure to these two DFT of order $N_1 = \frac{1}{2}N$ we have to compute four DFT of order $N_2 := \frac{1}{4}N$. This decomposition process has $\tau = \log_2 N$ stages. Since each stage requires $\frac{1}{2}N$ complex multiplications and N complex additions the number of arithmetic operations needed to compute the DFT of the data $z_0, \ldots, z_{N-1}$ is proportional to $N \log_2 N$.

<u>Remark 10.9</u>

When implementing a deconvolution method attention has to
be focused on the trade-off between resolution and accuracy.
A measure of resolution of a specific method can be defined
by specifying those objects (signals, pictures,...) which can
be seen as distinct. Commonly used objects are superpositions
of Dirac-functions; Lorentz-functions (see Ex. 10.1) approximate such objects numerically.   *

10.4 Reconstruction by successive approximation

Let T be a nonexpansive mapping in a Hilbert space X, i.e.
a linear operator with $\|T\| \leq 1$. We consider the algorithm

$$\text{Choose } x^{o} \in X, \text{ continue with}$$

$$(10.20) \qquad x^{k+1} = Tx^{k}, \quad k \in \mathbb{N}.$$

A special case we are interested in is given by

$$T := P_{m} \cdots P_{1} \quad \text{where } P_{i} \text{ is the orthogonal}$$
projection onto a closed
subspace N_{i}, $1 \leq i \leq m$.

The following two examples (reconstruction of a signal, reconstruction of a density) may be considered as a motivation
to investigate the algorithm above. Moreover, the ART-algorithm
with relaxation parameter $\lambda_{k}=1$ (see Section 9.5) can be considered as a special case of the iteration (10.20).

<u>Example 10.10</u>

Let $x \in L_{2}(\mathbb{R})$ and assume that we know the time segment
$y := x|_{[-T,T]}$, $T > 0$. Without any further knowledge about x we
cannot reconstruct x from y. But if we have the information
that the frequency spectrum of x is contained in $[-\Omega,\Omega]$, $\Omega > 0$,
which means $\hat{x} = \hat{x}\chi|_{[-\Omega,\Omega]}$, then, in principle, it is possible to

reconstruct x from y. This follows from the fact that x must be an analytic function due to the representation

$$x(t) = \frac{1}{\sqrt{2\pi}} \int_{-\Omega}^{\Omega} \hat{x}(\omega) e^{i\omega t} \, d\omega \ , \quad t \in \mathbb{R}.$$

If we define

$$D_T := \{f \in L_2(\mathbb{R}) \mid f\chi_{[-T,T]} = f\} \quad \text{(time-limited functions)},$$

$$B_\Omega := \{f \in L_2(\mathbb{R}) \mid \hat{f}\chi_{[-\Omega,\Omega]} = \hat{f}\} \quad \text{(band-limited functions)}$$

then D_T and B_Ω are closed subspaces of $L_2(\mathbb{R})$ and the ortho-gonal projections of $L_2(\mathbb{R})$ onto D_T and B_Ω exist ; they are of the following form:

$$D_T : L_2(\mathbb{R}) \longrightarrow D_T \ , \quad f \longmapsto f\chi_{[-T,T]} \ ,$$

$$B_\Omega : L_2(\mathbb{R}) \longrightarrow B_\Omega \ , \quad f \longmapsto \overset{\vee}{g} \text{ where } g = \hat{f}\chi_{[-\Omega,\Omega]} .$$

Then the reconstruction problem may be reformulated as follows:

Given $y \in D_T$ find $x \in L_2(\mathbb{R})$ with

$$(I - B_\Omega)x = \Theta \ , \quad D_T x = y. \hspace{3cm} *$$

Example 10.11

The reconstruction problem of tomography consists in deter-mining the density function f from a knowledge of certain radiographs $y_i := R_{\theta_i} f$, $1 \leq i \leq m$ (see Example (E3) in Ch. 1.) One can show that R_{θ_i} may be reasonable defined from a Hilbert space X into a Hilbert space Y, $1 \leq i \leq m$. Suppose that N_i is the null space of R_{θ_i} and that x^o is the true solution of

$$(*) \hspace{3cm} R_{\theta_i} x = y_i \ , \quad 1 \leq i \leq m.$$

Let P_i denote the projection of X onto N_i, $1 \leq i \leq m$. Then

problem $(*)$ is equivalent to the determination of x with
$x - x^o \in \bigcap_{i=1}^{m} N_i$. This is equivalent to

$$P_m \ldots P_1 (x - x^o) = x - x^o$$

as it is shown in Lemma 10.12. $\qquad\qquad *$

<u>Lemma 10.12</u>

Let $P_1, \ldots, P_m$ be orthogonal projections on a Hilbert space X and let $N_i = R(P_i)$, $1 \leq i \leq m$. Let $z \in X$. Then the following properties are equivalent:

$$\text{a)} \quad z \in \bigcap_{i=1}^{m} N_i \quad .$$

$$\text{b)} \quad P_m \ldots P_1 z = z .$$

<u>Proof:</u> The implication "a $\Rightarrow$ b" is immediately clear. To prove the converse implication "b $\Rightarrow$ a", let $z = z^1 + z^2$ with $z^1 \in N_1$, $z^2 \in N_1^{\perp}$. Then

$$\|z^1\|^2 + \|z^2\|^2 = \|z\|^2 = \|P_m \ldots P_1 z\|^2 = \|P_m \ldots P_2 z^1\|^2 \leq \|z^1\|^2$$

and therefore $z^2 = \theta$ and $z = z^1 \in N_1$. In the same way we obtain $z \in N_i$, $2 \leq i \leq m$. $\qquad\qquad \square$

Now, let us investigate the convergence properties of the algorithm (10.20). First, we need some properties of nonexpansive operators.

<u>Lemma 10.13</u>

We have

$$\text{1)} \quad \|T^*\| \leq 1 .$$

$$\text{2)} \quad N(I - T) = N(I - T^*) .$$

$$\text{3)} \quad X = N(I - T) \oplus \overline{R(I - T)} .$$

<u>Proof:</u>

Part 1) follows from

$$\|T^*x\|^2 = (T^*x,T^*x) = (TT^*x,x) \leq \|TT^*x\| \, \|x\| \leq \|T^*x\| \, \|x\|.$$

To prove 2) let $x \in N(I - T)$. Then

$$\|x\|^2 = (x,x) = (x,Tx) = (T^*x,x) \leq \|T^*x\| \, \|x\| \leq \|x\|^2$$

and therefore

$$(x,Tx) = \|x\| \, \|T^*x\|, \quad \|T^*x\| = \|x\|.$$

From

$$\|x - T^*x\|^2 = \|x\|^2 - 2(Tx,x) + \|T^*x\|^2 = 0$$

follows $x \in N(I - T^*)$. The converse inclusion follows similar.

Since $N(I - T^*) = R(I - T)^\perp$ we obtain from 2) $\overline{R(I - T)} = R(I - T)^{\perp\perp} = N(I - T^*)^\perp = N(I - T)^\perp$ and the projection theorem implies the result 3). $\qquad\square$

Let P_T denote the orthogonal projection of X onto $N(I - T)$.

<u>Lemma 10.14</u>

The following properties are equivalent:

a) $(T^k x)_{k \in \mathbb{N}}$ converges to $P_T x$ for all $x \in X$.

b) $(T^k(I - T)z)_{k \in \mathbb{N}}$ converges to zero for all $z \in X$.

c) $(T^k u)_{k \in \mathbb{N}}$ converges to zero for all $u \in \overline{R(I - T)}$.

d) $(T^k w)_{k \in \mathbb{N}}$ converges to zero for all $w \in R(I - T)$.

<u>Proof:</u> The implications "b) $\Rightarrow$ c)" and "c) $\Rightarrow$ d)" are clear.

a) $\Rightarrow$ b) Let $z \in X$. By a) $\lim_k T^k z = P_T z$. This implies

$$\lim_k T^{k+1}z - T^k z = \lim_k T^k (I - T)z = \theta.$$

d) $\Rightarrow$ c) Let $u \in \overline{R(I - T)}$ and let $\varepsilon > 0$. Then there exists $w \in R(I - T)$ with $\|w - u\| < \frac{\varepsilon}{2}$ and $k_o \in \mathbb{N}$ with $\|T^k w\| < \frac{\varepsilon}{2}$ for all $k \geq k_o$. Then

$$\|T^k u\| \leq \|T^k (u - w)\| + \|T^k w\| \leq \|u - w\| + \|T^k w\| < \varepsilon$$

for all $k \geq k_o$.

c) $\Rightarrow$ a) Let $x \in X$ and $x = v + u$ with $v = P_T x$ and $u \in \overline{R(I - T)}$ (see Lemma 10.13). Then

$$T^k x = T^k v + T^k u = v + T^k u, \quad \lim_k T^k x = v. \qquad \Box$$

Theorem 10.15

Suppose that the following property is satisfied:

If $(z_k)_{k \in \mathbb{N}}$ is a sequence in X with

$$(10.21) \quad \|z_k\| \leq 1, \ k \in \mathbb{N}, \ \lim_k \|Tz_k\| = 1$$

then $\lim_k (I - T)z_k = \theta$.

Then $(T^k x)_{k \in \mathbb{N}}$ converges to $P_T x$ for all $x \in X$.

Proof: We verify the condition b) in L.10.14. Let $z \in X$. Then $\|T^{k+1} z\| \leq \|T^k z\|$ for all $k \in \mathbb{N}$ and $b := \lim_k \|T^k z\|$ exists. If $b = 0$ then obviously $\lim_k T^k (I - T)z = \theta$. If $b > 0$ then

$$\|z_k\| \leq 1, \ k \in \mathbb{N}, \ \lim_k \|Tz_k\| = 1$$

for $z_k := T^k z \|T^k z\|^{-1}$, $k \in \mathbb{N}$. Therefore, by the assumption

$$\theta = \lim_k (I - T)z_k = \lim_k \|T^k z\|^{-1} (I - T)T^k z.$$

Since $\lim_k \|T^k z\|^{-1} = b^{-1} > 0$ we obtain

$$\lim_k (I - T)T^k z = \theta. \qquad \Box$$

In the following lemma we consider two situations in which the condition (10.21) is satisfied.

Lemma 10.16

a) If T is selfadjoint and nonnegative then the condition (10.21) is satisfied.

b) If $T = T_2 T_1$ where each T_i has the property (10.21) then T has the property (10.21).

Proof: The result a) follows from

$$\| (I - T) z \|^2 = ((I - T)^2 z, z) \leq ((I - T)(I + T) z, z) = \| z \|^2 - \| Tz \|^2.$$

To prove b) let $(z_k)_{k \in \mathbb{N}}$ be a sequence in X with

$$\| z_k \| \leq 1, \quad k \in \mathbb{N}, \quad \lim_k \| Tz_k \| = 1.$$

Then

$$\| T_1 z_k \| \leq 1, \quad k \in \mathbb{N}, \quad \lim_k \| T_1 z_k \| = 1 ,$$

so that

$$\lim_k (I - T) z_k = \lim_k (I - T_1) z_k + \lim_k (I - T_2) T_1 z_k = \theta.$$

$\square$

From now we consider the special case that the nonexpansive mapping T is given by $P_m \ldots P_1$ where each P_i is an orthogonal projection of X onto the closed subspace N_i.

Let $N := \bigcap_{i=1}^{m} N_i$ and let P be the orthogonal projection onto N. Since each P_i has property (10.21) $P_m \ldots P_1$ has property (10.21) (see Lemma 10.16). Therefore

$$\lim_k T^k x = Px$$

for each $x \in X$ by Th.10.15. In the following we want to give a bound for the rate of convergence of the iteration (10.20).

Definition 10.17

Let M_1 and M_2 be closed subspaces of a Hilbert space H with intersection M. The (acute) angle α between M_1 and M_2 is given by

$$\cos\alpha := \sup\left\{ \frac{(u,v)_H}{\|u\|_H \, \|v\|_H} \;\Big|\; u \in M_1 \cap M^\perp, \; v \in M_2 \cap M^\perp \right\}.$$

Theorem 10.18

If α_i is the angle between N_i and $\displaystyle\bigcap_{j=i+1}^{m} N_j$ then for any x in X

$$\lim_n T^k x = Px \; ,$$

$$\|T^k x - Px\| \le c^k \|x - Px\|, \; k \in \mathbb{N},$$

where

$$c^2 \le 1 - \prod_{j=1}^{m-1} \sin^2\alpha_j.$$

Proof: We have only to verify the bound for $\|T^k x - Px\|$. Let $x \in X$, $x = Px + (I - P)x$. Then $T^k Px = Px$ and

$$\|T^k x - Px\| = \|T^k Px + T^k (I - P)x - Px\|$$

$$= \|T^k (I - P)x\|$$

$$\le \left\| T\big|_{N^\perp} \right\|^k \|(I - P)x\|$$

$$= \left\| T\big|_{N^\perp} \right\|^k \|x - Px\|$$

for all $k \in \mathbb{N}$. Thus, it is sufficient to show that $1 - \displaystyle\prod_{j=1}^{m-1} \sin^2\alpha_j$ is an upper bound for $\left\| T\big|_{N^\perp} \right\|^2$. Let us consider first the case $m = 2$. If $u \in N_1 \cap N^\perp$ and $v \in N_2 \cap N^\perp$ then

$$\frac{(u,v)}{\|u\|\,\|v\|} = \frac{(u,P_2 v)}{\|u\|\,\|v\|} = \frac{(P_2 P_1 u, v)}{\|u\|\,\|v\|} \le \frac{\|P_2 P_1 u\|}{\|u\|} = \frac{\|P_2 u\|}{\|u\|}$$

and

$$\frac{(u,P_2u)}{\|u\|\ \|P_2u\|} = \frac{(P_2u,P_2u)}{\|u\|\ \|P_2u\|} = \frac{\|P_2u\|}{\|u\|} \ .$$

Hence

$$\cos\alpha_1 = \sup\left\{\frac{\|P_2u\|}{\|u\|}\ \Big|\ u \in N_1 \cap N^\perp\right\}\ .$$

Let $x \in N^\perp$, $x = P_1x + (I - P_1)x$. Then

$$\frac{\|Tx\|^2}{\|x\|^2} = \frac{\|P_2P_1x\|^2}{\|x\|^2} = \frac{\|P_2P_1x\|^2}{\|P_1x\|^2 + \|(I-P_1)x\|^2} \le \frac{\|P_2P_1x\|^2}{\|P_1x\|^2}\ .$$

From this we obtain

$$\left\|P_2P_1\Big|_{N^\perp}\right\|^2 \le \sup\left\{\frac{\|P_2u\|^2}{\|u\|^2}\ \Big|\ u \in N_1 \cap N^\perp\right\} = \cos^2\alpha_1 = 1 - \sin^2\alpha_1.$$

The result for $m \ge 2$ follows by induction as follows:

$$P_m\cdots P_1 = (P_mP_{m-1})(P_{m-1}\cdots P_1)$$

$$\left\|P_m\cdots P_1\Big|_{N^\perp}\right\|^2 \le \left\|P_mP_{m-1}\Big|_{N^\perp}\right\|^2 \left\|P_{m-1}\cdots P_1\Big|_{N^\perp}\right\|^2$$

$$\le (1 - \sin^2\alpha_{m-1})(1 - \prod_{j=1}^{m-2}\sin^2\alpha_j) \le 1 - \prod_{j=1}^{m-1}\sin^2\alpha_j\ .$$

$\square$

<u>Bibliographical comments</u>

An introduction into the theory of Fourier transforms can be found in [102]. The results in Section 10.2 are taken from AREF'EVA [5],[6]. The discretization of convolution equations is described in DAVIES [25]. A complete treatment of fast transforms is given in ELLIOT, RAO [32]. The reconstruction by successive approximation is based on von NEUMANN'S alternating-projection theorem [80]. The procedure can be applied in signal theory [7] [107] and computer tomography [48],[50].

Chapter 11: The final value problem

One of the classical ill-posed problems (in the sense of Hadamard) is the final value problem for evolution equations; a special case of this problem is the solution of the heat equation backwards in time. We consider some regularization methods for this ill-posed problem.

11.1 Introduction

Let $H := H_o$ be a separable Hilbert space with inner product $(.,.)_o$ and let $K : H \longrightarrow H$ be a bounded linear operator satisfying the following conditions:

$\quad$ i) $\quad$ K is compact with dim $R(K) = \infty$
$\quad\quad\quad$ and $R(K)$ dense in H;

(11.1)

$\quad$ ii) $\quad$ -K is selfadjoint and positive definite.

Then we know that

$$A := K^{-1} : R(K) \longrightarrow H$$

exists. We may A consider as an unbounded operator from H into H with $D(A) := R(K)$ as its domain of definition. With these notations let us consider the following initial value problems:

$\quad$ Given $T > O$ and $v \in H$, find a mapping
$\quad$ $z : [O,T] \longrightarrow H$ with

$\quad\quad$ i) $\quad$ $z(t) \in D(A)$ $\quad$, $t \in (O,T)$.

(11.2) $\quad$ ii) $\quad$ $z(o) = v$.

$\quad\quad$ iii) $\quad$ $\dot{z}(t) = Az(t)$ $\quad$, $t \in (O,T)$.

Here the dot indicates derivative with respect to the "time-variable" t; the exact meaning becomes clear in the sequel.

The problem which is the subject of this chapter results from
(11.2) by changing the initial condition ii) into a final
condition:

$$\text{Given } T > 0 \text{ and } u \in H, \text{ find a mapping}$$
$$z : [0,T] \longrightarrow H \text{ such that}$$

$$\text{i)} \quad z(t) \in D(A) \quad , \quad t \in (0,T) \ .$$

(11.3)
$$\text{ii)} \quad z(T) = u \ .$$

$$\text{iii)} \quad \dot{z}(t) = Az(t) \ , \ t \in (0,T).$$

We call this problem a final value problem. In the following
example we consider the simplest problem of a final value
problem, the heat equation backwards in time.

Example 11.1

Let $H := L_2[0,1]$ with the usual inner product and define
$K : H \longrightarrow H$ by

$$(Kz)(x) := \int_0^1 \kappa(x,s)z(s)\,ds$$

where

$$\kappa(x,s) := \begin{cases} x(1-s) & , \ 0 \le x \le s \le 1 \\ s(1-x) & , \ 0 \le s \le x \le 1 \end{cases}.$$

Clearly, $(Kz)(0) = (Kz)(1) = 0$, $z = (Kz)'' \in L_2[0,1]$, and
therefore

$$D(A) = \{z \in H^2[0,1]\,|\,z(0) = z(1) = 0\},$$

$$Az = z'' \ , \ z \in D(A).$$

The problem (11.3) consists in solving the heat equation backwards in time:

$$z_t = z_{xx} \qquad\qquad , \quad 0 < x < 1, \ 0 < t < T,$$

$$z(t,0) = z(t,1) = 0 \ , \quad 0 < t < T,$$

$$z(T,x) = u(x) \qquad\qquad , \quad 0 < x < 1.$$

Since the solution of the heat equation is smooth for times
$t > 0$ (see Th. 11.7) the function u must be smooth.

Physically this problem can be viewed as the problem of finding the temperature distribution for times $t < T$ of a thin rod whose ends at $x = 0$ and $x = 1$ are kept at $0^\circ C$, given a knowledge of the temperature distribution in the rod at time $t = T$. Since heat conduction is a irreversible phenomenon the reconstruction of the past $(t < T)$ from the present $(t = T)$ leads to tremendous difficulties.   *

In the sequel we assume without further mention that the condition (11.1) is satisfied and that A is given as K^{-1}.

Lemma 11.2

There exists a sequence $(\lambda_n)_{n \in I\!N}$ and an orthonormal basis $(e_n)_{n \in I\!N}$ of H such that

i) $0 < \lambda_n \leq \lambda_{n+1}$, $n \in I\!N$; $\lim\limits_{n} \lambda_n = \infty$;

ii) $e_n \in D(A)$, $Ae_n + \lambda_n e_n = 0$, $n \in I\!N$;

iii) $v \in D(A)$ if and only if $\sum\limits_{n \in I\!N} |(v, e_n)_o|^2 \lambda_n^2 < \infty$.

Proof: By assumption (11.1) there exists a sequence $(\mu_n)_{n \in I\!N}$ and an orthonormal basis of H satisfying

1) $0 < \mu_{n+1} \leq \mu_n$, $n \in I\!N$, $\lim\limits_{n} \mu_n = 0$;

2) $Ke_n = \mu_n e_n$, $n \in I\!N$;

3) $v \in R(K)$ if and only if $\sum\limits_{n \in I\!N} |(v, e_n)_o|^2 \mu_n^{-2} < \infty$

(see Section 4.3). Setting $\lambda_n := \mu_n^{-1}$ the assertions follow .

$\square$

Examples 11.3

   1) In the setting of Example 11.1 we have

$$\lambda_n = (\pi n)^2, \quad e_n(x) = \sqrt{2} \sin \sqrt{\lambda_n}x, \quad x \in [0,1]; \quad n \in \mathbb{N}.$$

   2) Let us consider the diffusion equation in two dimensions with periodic boundary conditions:

$$\begin{cases} z_t = \alpha(z_{xx} + z_{yy}), \text{ in } (0,2\pi) \times (0,2\pi) , \\[2ex] z(t,0,y) = z(t,2\pi,y) \quad , \quad z_x(t,0,y) = z_x(t,2\pi,y), \ 0 < y < 2\pi, \\[2ex] z(t,x,0) = z(t,x,2\pi) \quad , \quad z_y(t,x,0) = z_y(t,x,2\pi), \ 0 < x < 2\pi. \end{cases}$$

With the condition

$$z(T,x,y) = u(x,y) \qquad , \quad 0 < x < 2\pi, \ 0 < y < 2\pi$$

at the final time $T > 0$ we have again a problem of the type (11.3) if we define A by

$$(Az)(t) := \alpha \sum_{m,n \in \mathbb{N}} (z(t),e_{mn})_o e_{mn} \lambda_{mn}$$

where

$$\lambda_{mn} = m^2 + n^2, \quad e_{mn}(x,y) = \frac{1}{2\pi} e^{imx+iny}, \quad m,n \in \mathbb{Z},$$

and $(.,.)_o$ is the usual inner product in $L_2([0,2\pi] \times [0,2\pi])$ (with an obvious interpretation in the real Hilbert space setting).

   3) Using the sequences $(e_{mn})_{m,n \in \mathbb{Z}}$ and $(\lambda_{mn})_{m,n \in \mathbb{Z}}$ of Example 2), we may define

$$(Az)(t) := \alpha \sum_{m,n \in \mathbb{Z}} (z(t),e_{mn})_o \ e_{mn} \ \lambda_{mn}^{5/6}.$$

(The resulting problem (11.3) describes a reconstruction problem in image processing).   *

The superposition principle suggests the following ansatz for the solution of the problem (11.3):

$$z(t) = \sum_{n \in \mathbb{N}} c_n(t) e_n \ , \ t > 0.$$

By formal computations the conditions ii),iii) lead us to the following family of (scalar) final value problems:

$$(11.4) \qquad \begin{cases} c_n'(t) = - \lambda_n c_n(t) \ , \ t \in (0,T) \\ c_n(T) = (u,e_n)_o \end{cases} .$$

The unique solution is given by

$$c_n(t) = e^{-\lambda_n (t-T)} (u,e_n)_o \ , n \in \mathbb{N} .$$

Therefore a candidate for the solution of (11.3) is given by

$$(11.5) \qquad z(t) = \sum_{n \in \mathbb{N}} e^{-\lambda_n (t-T)} (u,e_n)_o e_n .$$

With this formula for the solution we see immediately that the final value problem (11.3) is ill-posed: For example, for $u := u_k := e_k$ the solution $z = z_k$ is given by

$$z_k(t) := e^{-\lambda_k (t-T)} e_k, \ t > 0,$$

and we have

$$\lim_k \| u_k \|_o = 1, \ \lim_k \| z_k(t) \|_o = \infty \ \text{for every } t \in [0,T).$$

11.2 The mild solution of the forward problem

In this section we consider the initial value problem (11.2) in order to establish helpful results for studying the inverse problem (11.3).

To begin with, let us introduce formally a mapping G on H by

$$(11.6) \qquad G(t)v := \sum_{n \in \mathbb{N}} e^{-\lambda_n t} (v,e_n)_o e_n \ , \ t \geq 0$$

(see formula (11.5)).

Lemma 11.4

We have

   i) $G(t) \in B(H)$, $\|G(t)\| \le 1$, $t \ge 0$;

  ii) $G(t)G(s) = G(t+s)$, $t,s \ge 0$;

 iii) $G(0) = I$;

  iv) $[0,\infty) \ni t \longmapsto G(t)v \in H$ is continuous for all $v \in H$.

Proof: The linearity of G is obvious.

$$|G(t)v|_o^2 = \sum_{n \in \mathbb{N}} e^{-2\lambda_n t} |(v,e_n)_o|^2 \le \sum_{n \in \mathbb{N}} |(v,e_n)|^2 = \|v\|_o^2 .$$

This proves part i). The assertions ii) and iii) are immediate consequences of the definition of G. The proof of part iv) goes as follows: Let $\varepsilon > 0$ and let $N \in \mathbb{N}$ such that

$$\sum_{n \ge N+1} |(v,e_n)_o|^2 \le \varepsilon^2 .$$

Choose $\delta > 0$ such that for all t,t' with $|t - t'| < \delta$

$$|e^{-\lambda_n t} - e^{-\lambda_n t'}| \le \varepsilon \text{ for all } n \le N.$$

Then we have for $t,t' \ge 0$ with $|t - t'| < \delta$:

$$|G(t)v - G(t')v\|_o^2 = \sum_{n=1}^{N} |(v,e_n)_o|^2 (e^{-\lambda_n t} - e^{-\lambda_n t'})^2$$

$$+ \sum_{n \ge N+1} |(v,e_n)_o|^2 (e^{-\lambda_n t} - e^{-\lambda_n t'})^2$$

$$\le \varepsilon^2 \sum_{n=1}^{N} |(v,e_n)_o|^2 + \sum_{n \ge N+1} |(v,e_n)_o|^2 \cdot 4 \le \varepsilon^2 \|v\|_o^2 + 4\varepsilon^2 .$$

$\square$

Lemma 11.5

We have for each $v \in D(A)$

$$\lim_{h \searrow 0} h^{-1} (G(h)v - v) = Av .$$

<u>Proof:</u> Let $\varepsilon > 0$ and choose $N \in \mathbb{N}$ with

$$\sum_{n \geq N+1} |(v,e_n)_o|^2 \, \lambda_n^2 < \varepsilon^2.$$

This choice is possible since $v \in D(A)$ (see Lemma 11.2).

Choose $h_o > 0$ such that

$$h^{-1} |e^{-\lambda_n h} - 1 + \lambda_n h| \leq \varepsilon \quad \text{for all } h \in (0, h_o] \text{ and } n \leq \mathbb{N}.$$

Since

$$|e^{-s} - 1 + s| \leq 2s \quad \text{for all } s \geq 0$$

we obtain for $h \in (0, h_o]$

$$h^{-2} \| G(h)v - v - hAv \|_o^2$$

$$\leq \sum_{n \in \mathbb{N}} |(v,e_n)_o|^2 (e^{-\lambda_n h} - 1 + \lambda_n h)^2 h^{-2}$$

$$\leq \varepsilon^2 \sum_{n=1}^{N} |(v,e_n)_o|^2 + 4 \sum_{n \geq N+1} |(v,e_n)_o|^2 \, \lambda_n^2$$

$$\leq \varepsilon^2 \| v \|_o^2 + 4\varepsilon^2 . \qquad\qquad \square$$

<u>Remark 11.6</u>

The results of L.11.4 and L.11.5 show that the family $(G(t))_{t \geq o}$ is a semigroup with infinitesimal generator A.

*

<u>Theorem 11.7</u>

Let $v \in D(A)$ and define $w(t) := G(t)v$, $t \geq 0$.
Then

   i) $w(t) \in D(A)$, $t \geq 0$.

  ii) $(0,\infty) \ni t \longmapsto w(t) \in H$ is differentiable with
      derivative $\dot{w}$.

 iii) $\dot{w}(t) - Aw(t) = G(t)Av$, $t > 0$.

  iv) $w(0) = v$.

__Proof:__ Part i) follows with Lemma 11.2 from

$$\sum_{n \in \mathbb{N}} |(w(t),e_n)_o|^2 \lambda_n^2 = \sum_{n \in \mathbb{N}} |(v,e_n)_o|^2 e^{-2\lambda_n t} \lambda_n^2$$

$$\leq \sum_{n \in \mathbb{N}} |(v,e_n)_o|^2 \lambda_n^2 < \infty.$$

For h > O we have by Lemma 11.4

$$w(t + h) - w(t) = G(t)(G(h)v - v) ,$$

$$w(t - h) - w(t) = G(t - h)(v - G(h)v) .$$

Using L.11.5 we see immediately that the assertions ii) and iii) are valid. The identity iv) is clear.

$\square$

The last theorem shows that the formula (11.6) may be used to define a solution of the problem (11.2). This is the content of the following

Definition 11.8

For every $v \in H$ we call $w(\cdot) := G(\cdot)v$ the __mild solution__ of the problem (11.2).

As a consequence of the results above we see that the forward problem (11.2) is well-posed if we use the solution concept of Def.11.8; the continuous dependence of the mild solution on the initial value follows from $\|G(t)\| \leq 1$, t > O.

With the family $(G(t))_{t \geq 0}$ we may reformulate the final value problem (11.3) as follows:

(11.7)   Given T > O and $u \in H$, find $v \in H$ such that

$$G(T)v = u .$$

This problem is ill-posed as the formal argument at the end of Section 11.1 shows. In other words, the solution of (11.7) doesn't exist for every u, and on the set of the u's for which it does exist, the solution does not depend continuously on u. So we have to regularize the problem. This is the content of Section 11.4. In the next section we establish results which characterize the domain of definition of $G(T)^{-1}$.

11.3 The Hilbert scales $E_{a,t}$

For each $t \geq 0$ the sequence $(e^{-\lambda_n t} e_n)_{n \in \mathbb{N}}$ is an orthogonal sequence whose linear hull is dense in H; moreover the sequence $(\lambda_n^{-1})_{n \in \mathbb{N}}$ satisfies

$$0 < \lambda_{n+1}^{-1} \leq \lambda_n^{-1}, \quad n \in \mathbb{N}, \quad \lim_n \lambda_n^{-1} = 0.$$

Therefore, by Section 5.1, we have for each $t \geq 0$ a Hilbert scale

$$(E_{a,t}, \ (\cdot,\cdot)_{a,t})_{a \in \mathbb{R}}$$

where for each $a \geq 0$

$$E_{a,t} = \{w \in H \mid \sum_{n \in \mathbb{N}} |(w,e_n)_0|^2 \, e^{-2\lambda_n t} \lambda_n^{2a} < \infty\}$$

$$(w,\tilde{w})_{a,t} = \sum_{n \in \mathbb{N}} (w,e_n)_0 (\tilde{w},e_n)_0 \lambda_n^{2a} e^{-2\lambda_n t}, \quad w,\tilde{w} \in E_{a,t}.$$

Corollary 11.9

i) $E_{0,0} = H$, $E_{1,0} = D(A)$;

ii) $E_{a,t} \subset E_{b,t}$ for all $a,b \in \mathbb{R}$ if $0 \leq t \leq s$;

iii) $E_{a,t} \subset E_{b,s}$ if $a \leq b$ and $0 \leq t \leq s$;

iv) $E_{a,t}$ is dense in $E_{b,s}$ if $a \leq b$ and $0 \leq t \leq s$.

Proof: It is enough to proof the assertion for spaces $E_{a,t}$, $a \geq 0$, due the duality relation

$$E_{a,t} = E_{-a,t}^*.$$

i) follows from L.11.2. To prove ii) let $0 \leq t \leq s$, $0 \leq a,b$, and let $w \in E_{a,t}$. Then

$$\sum_{n \in \mathbb{N}} |(w,e_n)_o|^2 \lambda_n^{2b} e^{-2\lambda_n s}$$

$$= \sum_{n \in \mathbb{N}} |(w,e_n)_o|^2 \lambda_n^{2a} e^{-2\lambda_n t} \lambda_n^{2b-2a} e^{-2\lambda_n (t-s)}$$

$$\leq \sup_{\lambda \geq \lambda_1} (\lambda^{2b-2a} e^{-2\lambda(s-t)}) \|w\|_{a,t}^2 < \infty \quad .$$

Part iii) is proved similar to part ii). We know that

$$M := \{w \in H \mid (w,e_n)_o = 0 \text{ for almost all } n \in \mathbb{N}\}$$

is dense in $E_{b,s}$. Since $M \subset E_{a,s} \subset E_{b,s}$ we obtain the result iv).

$$\Box$$

Corollary 11.10

1) We have for each $a \in \mathbb{R}$ and all r,t,s with $0 \leq r \leq t \leq s,\ s \neq r,$

$$\|w\|_{a,t} \leq \|w\|_{a,r}^{\theta} \|w\|_{a,s}^{1-\theta} , \quad w \in E_{a,r} ,$$

where $\theta = \dfrac{s-t}{s-r}$.

2) We have for each $t \geq 0$ and all $a,b,c \in \mathbb{R}$ with $a \leq b \leq c,\ c \neq a,$

$$\|w\|_{b,t} \leq \|w\|_{a,t}^{\theta} \|w\|_{c,t}^{1-\theta} , \quad w \in E_{c,t} ,$$

where $\theta = \dfrac{c-b}{c-a}$.

Proof: The result 2) follows from Theorem 5.3. We prove 1).

$$\|w\|_{a,t}^2 = \sum_{n \in \mathbb{N}} |(w,e_n)_o|^2 \lambda_n^{2a} e^{-2\lambda_n t}$$

$$= \sum_{n \in \mathbb{N}} (|(w,e_n)_o|^2 \lambda_n^{2a} e^{-2\lambda_n r})^{\theta} (|(w,e_n)_o|^2 \lambda_n^{2a} e^{-2\lambda_n s})^{1-\theta}$$

$$\leq \left(\sum_{n\in\mathbb{N}} |(w,e_n)_o|^2 \lambda_n^{2a} e^{-2\lambda_n^r} \right)^{\theta} \left(\sum_{n\in\mathbb{N}} |(w,e_n)_o|^2 \lambda_n^{2a} e^{-2\lambda_n^s} \right)^{1-\theta}$$

$$= \|w\|_{a,r}^{2\theta} \, \|w\|_{a,s}^{2(1-\theta)} . \qquad \square$$

<u>Corollary 11.11</u>

The family $(\lambda_n^{-a} e^{-\lambda_n t} e_n)_{n\in\mathbb{N}}$ is an orthonormal basis in $E_{a,t}$ for each $a \in \mathbb{R}$ and $t \geq 0$.

<u>Proof</u>: By construction of $E_{a,t}$. $\qquad \square$

<u>Lemma 11.12</u>

1) If $t < s$ and $a,b \in \mathbb{R}$ then the imbedding

$$J : E_{a,t} \longrightarrow E_{b,t}$$

is compact.

2) If $a > b$ then the imbedding

$$J : E_{a,t} \longrightarrow E_{b,t}$$

is compact.

<u>Proof</u>: We prove only 1), the proof of 2) is similar. We know $E_{a,t} \subset E_{b,s}$. Let $w \in E_{a,t}$. Then we have by Corollary 11.11

$$w = \sum_{n\in\mathbb{N}} (w, \lambda_n^a e^{\lambda_n t} e_n)_{a,t} \, \lambda_n^{-a} e^{-\lambda_n t} e_n$$

$$= \sum_{n\in\mathbb{N}} \lambda_n^{b-a} e^{\lambda_n(t-s)} (w, \lambda_n^a e^{\lambda_n t} e_n)_{a,t} \, \lambda_n^b e^{\lambda_n s} e_n.$$

Since $(\lambda_n^{b-a} e^{\lambda_n(t-s)})_{n\in\mathbb{N}}$ converges to zero if n goes to infinity the imbedding J is compact by Lemma 4.12.

$\qquad \square$

<u>Theorem 11.13</u>

Let $T > 0$, $v \in H$, and let z be the mild solution of the
initial value problem

$$\dot{z} = Az \ , \quad z(0) = v.$$

Then

$$(11.8) \qquad \|z(t)\|_o \leq \|z(0)\|_o^{1-\frac{t}{T}} \|z(T)\|_o^{\frac{t}{T}} \ , \quad t \in [0,T].$$

<u>Proof:</u> Apply Cor.11.10, 1) with $r = 0$, $s = T$, $a = 0$ and
notice that $\|z(t)\|_o = \|v\|_{o,t}$ by (11.6), the definition of the
mild solution and the construction of $\|\cdot\|_{o,t}$.

$\square$

<u>Remarks 11.14</u>

1) The inequality (11.8) is sharp since we have equality
whenever v has the form e_n.

2) The inequality (11.8) is called a logarithmic convexity
inequality due to the fact that

$$\ln\|z(t)\|_o \leq (1 - \tfrac{t}{T})\ln\|z(0)\|_o + \tfrac{t}{T}\ln\|z(T)\|_o.$$

This convexity result is the basis for a stabilization of the
ill-posed final value problem (11.3).   *

<u>11.4 Regularizing schemes</u>

Now we want to consider the final value problem under the
influence of error in the data. In the notation of the preceding sections we have to study the following problem:

Given $v^o, u^o, u^\varepsilon \in H$ with

$$(11.9) \qquad G(T)v^o = u^o, \quad \|u^o - u^\varepsilon\|_o \leq \varepsilon \ .$$

Reconstruct the initial value v^o from u^ε.

If v^ε is a reconstruction of v^o from the data u^ε we obtain a reconstruction $z^\varepsilon := G(\cdot)v^\varepsilon$ of the state $z^O := G(\cdot)v^O$ and we are interested in an estimate of $\|z^\varepsilon(t) - z^O(t)\|_O$, $t \in [0,T]$, in terms of ε. Since the final value problem is ill-posed we have to introduce some sort of a-priori information. As we shall see in the following theorem, a bound

$$(11.10) \qquad \|v^O\|_O \leq E$$

is sufficient for the stabilization of the problem.

Theorem 11.15

Suppose that $\|v^O\|_O \leq E$. Let $v^\varepsilon \in H$ with $\|v^\varepsilon\|_O \leq E$ and $\|G(T)v^\varepsilon - u^\varepsilon\|_O \leq \varepsilon$. Then

$$(11.11) \qquad \|G(t)v^\varepsilon - G(t)v^O\|_O \leq 2E^{1-\frac{t}{T}}\varepsilon^{\frac{t}{T}}, \quad t \in [0,T].$$

Proof: Let $z(t) := G(t)v^\varepsilon - G(t)v^O = G(t)(v^\varepsilon - v^O)$, $t \in [0,T]$. Then

$$\|z(0)\|_O \leq 2E, \quad \|z(T)\| \leq 2\varepsilon.$$

Inserting this on the right-hand side of (11.8) we obtain (11.11).

$\square$

Remarks 11.16

1) The inequality (11.11) provides at $t = 0$ only the (redundant) information that

$$\|v^\varepsilon - v^O\| \leq 2E.$$

2) The inequality (11.11) is the best possible result as long as no further information is applied a-priori concerning the noise or the initial value of the unknown state. This is a consequence of the sharpness of the logarithmic convexity result.

$*$

To find concrete reconstructions v^ε one may use any of the standard methods to solve ill-posed problems. We start here

from the formal solution (11.5) of the final value problem ,
using as final value the given data u^ε:

$$\sum_{n\in I\!N} e^{-\lambda_n(t-T)} (u^\varepsilon,e_n)_o \, e_n.$$

Clearly, this series cannot be used as a reconstruction
because it doesn't converge due to the fact that the sequence
$(e^{-\lambda_n(t-T)})_{n\in I\!N}$ is not bounded for $t < T$. Thus, we introduce
a damping function in the same manner as in Section 5.2.

<u>Definition 11.17</u>

A function $q : [0,T] \times (0,\infty) \times (0,\infty) \to I\!R$ is called a <u>damping</u>
<u>function</u> if and only if q satisfies the following condi-
tions:

i) $0 \leq q(t,\alpha,\lambda) \leq 1$, $(t,\alpha,\lambda) \in [0,T] \times (0,\infty) \times (0,\infty)$.

ii)   There exist for each $(t,\alpha) \in [0,T] \times (0,\infty)$
constants $\beta(t,\alpha)$, $\gamma(t,\alpha)$ such that

$$\sup_{\lambda>o} q(t,\alpha,\lambda)e^{-\lambda(t-T)} \leq \beta(t,\alpha),$$

$$\sup_{\lambda>o} e^{-\lambda t}(1 - q(t,\alpha,\lambda)) \leq \gamma(t,\alpha).$$

With such a damping function there is associated a family
$(R_\alpha(t))_{\alpha>o,\,t\in[0,T]}$ of reconstruction operators defined by

$$(11.12) \quad R_\alpha(t)u := \sum_{n\in I\!N} q(t,\alpha,\lambda_n)e^{-\lambda_n(t-T)} (u,e_n)_o e_n, t \in [0,T].$$

The properties in Def.11.17 become clear if we consider the
reconstruction error

$$\|R_\alpha(t)u^\varepsilon - G(t)v^o\|_o \,, \alpha > 0, \, t \in [0,T].$$

<u>Lemma 11.18</u>

Let the family $(R_\alpha(t))_{\alpha>0,\,t\in[0,T]}$ be defined by (11.12).
Then if $\|v^o\|_o \leq E$ we have

i) $R_\alpha(t) \in B(H)$,$\alpha > 0$, $t \in [0,T]$.

ii) $\|R_\alpha(t)u^\varepsilon - G(t)v^o\|_o \leq \beta(t,\alpha)\varepsilon + \gamma(t,\alpha)E$.

<u>Proof:</u> Let $u \in H$, $\alpha > 0$ and $t \in [0,T]$. Then

$$\|R_\alpha(t)u\|_o^2 = \sum_{n\in\mathbb{N}} |q(t,\alpha,\lambda_n)|^2 e^{-2\lambda_n(t-T)} |(u,e_n)_o|^2$$

$$\leq \sum_{n\in\mathbb{N}} \beta(t,\alpha)^2 |(u,e_n)_o|^2$$

$$= \beta(t,\alpha)^2 \|u\|_o^2 \quad .$$

This proves part i). The proof of part ii) starts from

$$\|R_\alpha(t)u^\varepsilon - G(t)v^o\|_o \leq \|R_\alpha(t)u^\varepsilon - R_\alpha(t)u^o\|_o + \|R_\alpha(t)u^o - G(t)v^o\|_o.$$

$$\|R_\alpha(t)u^\varepsilon - R_\alpha(t)u^o\|_o^2 = \sum_{n\in\mathbb{N}} |q(t,\alpha,\lambda_n)|^2 e^{-2\lambda_n(t-T)} |(u^\varepsilon-u^o,e_n)_o|^2$$

$$\leq \beta(t,\alpha)^2 \sum_{n\in\mathbb{N}} |(u^\varepsilon-u^o,e_n)_o|^2 = \beta(t,\alpha)^2 \|u^\varepsilon-u^o\|_o^2 \quad .$$

$$\|R_\alpha(t)u^o - G(t)v^o\|_o^2 = \|\sum_{n\in\mathbb{N}} (q(t,\alpha,\lambda_n)-1)e^{-\lambda_n(t-T)} (u^o,e_n)_o e_n\|_o^2$$

$$= \|\sum_{n\in\mathbb{N}} (q(t,\alpha,\lambda_n)-1)e^{-\lambda_n t} (v^o,e_n)_o e_n\|^2$$

$$\leq \gamma(t,\alpha)^2 \sum_{n\in\mathbb{N}} |(v^o,e_n)_o|^2 = \gamma(t,\alpha)^2 \|v^o\|_o^2. \qquad \square$$

Clearly, in order to come to an "optimal" reconstruction we minimize the bound

$$\eta(t,\alpha) := \beta(t,\alpha)\varepsilon + \gamma(t,\alpha)E$$

with respect to the regularizing parameter $\alpha > 0$.

Definition 11.19

Let the regularizing family $(R_\alpha(t))_{\alpha>0, t\in[0,T]}$ be associated with the damping function q. Then we call the resulting regularization method <u>optimal</u> if and only if there exist $\alpha^* > 0$ and $c \geqq 0$ such that

$$\|R_{\alpha^*}(t)u^\varepsilon - G(t)v^o\| \leq c\, E^{1-\frac{t}{T}}\, \varepsilon^{\frac{t}{T}} \quad \text{for all } t \in [0,T].$$

In Table 3 (see the following page) we give concrete examples of damping functions; moreover, we indicate for each regularization scheme that parameter which minimizes $\eta(t,\alpha)$ with respect to α.

In the calculation of $R_\alpha(T)u^\varepsilon$ via formula (11.12) one has to know the eigenvalues and eigenvectors of A. But in the case of the methods (3)-(5) (Table 5) one can see that the calculation of $R_\alpha(t)u^\varepsilon$ is equivalent to the solution of a differential equation which is different from the equation

$$\dot{z} = Az \ .$$

One can see this by the same arguments that led us to the formula (11.5).

<u>Ad(3):</u> $R_\alpha(t)u^\varepsilon = z^\varepsilon(t)$ where z^ε solves

$$(I - \alpha^{-1}A)\dot{z} = Az \ , \quad z(T) = u^\varepsilon.$$

(The equation for z^ε is a pseudoparabolic equation).

<u>Ad(4):</u> $R_\alpha(t)u^\varepsilon = z^\varepsilon(t)$ where z^ε solves

$$\dot{z} = (I + \alpha^{-1}A)Az \ , \quad z(T) = u^\varepsilon.$$

(Method of quasi-réversibilité).

<u>Ad(5):</u> $R_\alpha(t)u^\varepsilon = z^\varepsilon(t)$ where z solves

$$\ddot{z} = (A + \alpha I)^2 z, \quad z(0) = \Theta, \quad z(T) = e^{\alpha T}u^\varepsilon \ .$$

(Backward beam method. This equation results from the following observation: If z is a solution of $\dot{z} = Az$ then $\tilde{z} := e^{\alpha \cdot}z$ solves $\dot{z} = (\alpha I + A)z$. Differentiating with respect to time we get that $\tilde{z}$ solves the problem above).

Method	Damping function	$\beta(t,\alpha)$	$\gamma(t,\alpha)$	$\tilde{\alpha}$	$\eta(t,\tilde{\alpha})$
(1)	$q(t,\alpha,\lambda) := \begin{cases} 1 & ,\ \lambda \lesseqgtr \alpha \\ 0 & ,\ \lambda > \alpha \end{cases}$	$e^{\alpha(T-t)}$	$e^{-\alpha t}$	$\frac{1}{T}\ln(\frac{E}{\varepsilon})$	$2E^{\,1-\frac{t}{T}}\varepsilon^{\frac{t}{T}}$
(2)	$q(t,\alpha,\lambda) := \begin{cases} 1 & ,\ \lambda \leqq \alpha \\ e^{(\lambda-\alpha)(t-T)} & ,\ \lambda > \alpha \end{cases}$	$e^{\alpha(T-t)}$	$e^{-\alpha t}$	$\frac{1}{T}\ln(\frac{E}{\varepsilon})$	$2E^{\,1-\frac{t}{T}}\varepsilon^{\frac{t}{T}}$
(3)	$q(t,\alpha,\lambda) := e^{\frac{\lambda^2}{\alpha+\lambda}(t-T)}$	$e^{\alpha(T-t)}$	$(T-t)\dfrac{4}{t^2\alpha}$	$\frac{1}{T}\ln(\frac{E}{\varepsilon})$	$O((\ln(\frac{E}{\varepsilon}))^{-1})$
(4)	$q(t,\alpha,\lambda) := e^{\frac{\lambda^2}{\alpha}(t-T)}$	$e^{\frac{\alpha}{4}(T-t)}$	$(T-t)\dfrac{4}{t^2\alpha}$	$\frac{1}{T}\ln(\frac{E}{\varepsilon})$	$O((\ln(\frac{E}{\varepsilon}))^{-1})$
(5)	$q(t,\alpha,\lambda) :=$ $e^{(\alpha-\lambda)(T-t)}\,\dfrac{\sinh(\alpha-\lambda)t}{\sinh(\alpha-\lambda)T}$	$\frac{t}{T}e^{\alpha(T-t)}$	$\dfrac{T-t}{t}e^{-\alpha t}$	$\frac{1}{T}\ln(\frac{E}{\varepsilon})$	$E^{\,1-\frac{t}{T}}\varepsilon^{\frac{t}{T}}$
(6)	$q(t,\alpha,\lambda) :=$ $e^{(\alpha-\lambda)(T-t)}\,\dfrac{1+e^{(\alpha-\lambda)t}}{1+e^{(\alpha-\lambda)T}}$	$2e^{\alpha(T-t)}$	$e^{-\alpha t}$	$\frac{1}{T}\ln(\frac{E}{\varepsilon})$	$3E^{\,1-\frac{t}{T}}\varepsilon^{\frac{t}{T}}$
(7)	$q(t,\alpha,\lambda) := \dfrac{e^{-\lambda T}}{\alpha(\frac{T}{t}-1)+e^{-\lambda T}}$	$\frac{t}{T}(\frac{1}{\alpha})^{1-\frac{t}{T}}$	$(1-\frac{t}{T})\alpha^{\frac{t}{T}}$	$\frac{\varepsilon}{E}$	$E^{\,1-\frac{t}{T}}\varepsilon^{\frac{t}{T}}$

Table 3

$\underline{\text{Ad}(6)}$: $R_\alpha(t)u^\varepsilon = z^\varepsilon(t) + e^{\alpha(T-t)}(u^\varepsilon - z^\varepsilon(T))$ where z^ε solves

$$\dot{z} = Az, \quad z(0) + e^{\alpha T}z(T) = e^{\alpha T}u^\varepsilon.$$

We see from the table that the methods (5) and (6) yield optimal estimates. But these methods have some computational disadvantage: The problems which define z^ε are boundary value problems (in time).

<u>Remarks 11.20</u>

1) New regularizing schemes can be easily constructed by approximating exponential terms which are essential in the formulae above by simpler functions; for example

$$e^{-r} \approx (1 + r)^{-1} \ , \ e^{-r} \approx (1 - \tfrac{r}{2})(1 + \tfrac{r}{2})^{-1} \quad .$$

2) The numerical solution of a well-posed initial value problem is usually computed by a marching procedure. Clearly, such marching procedures would be interesting also for the final value problem. But, since this problem is ill-posed, if any of the standard marching finite difference schemes for solving the forward equation is implemented, with the time direction reversed, the resulting computation is unstable and will blow up as the mesh is refined.

3) If one approximates the exponential terms in the damping function (7) by Padé approximations one can achieve optimal error bounds and can see that the resulting regularizing scheme may be interpreted as a marching procedure. For example, the application of the Crank-Nicolson scheme backwards in time is associated with the approximation

$$e^{-r} \approx (1 - \tfrac{r}{2})(1 + \tfrac{r}{2})^{-1} \ . \qquad *$$

<u>Bibliographical comments</u>

A first reference concerning the heat equation backwards in time is DOETSCH [28]. The presentation in the first three sections follows mainly GRENACHER [42]. Our discussion of regularizing schemes is drawn from papers by BUZBEE, CARASSO [16], HÖHN [54], ELDEN [30] and ANG [3]. An application in imaging processing is described in CARASSO, SANDERSON, HYMAN [18].

Chapter 12: Parameteridentification

Models of systems in form of differential equations commonly contain parameters whose values are not fully known (see example (E5) in Chapter 1). In this chapter we consider some aspects of the identification of these model parameters from input-output data.

12.1 Identifiability of parameters in dynamical systems

Consider a differential system with observation equation

$$z' = f(t,z,u;p) \quad , \quad t \in [0,T]$$

(12.1) $\qquad z(0) = z_o \in Z_o, \quad u \in U, \quad p \in Q$

$$y = g(z;p)$$

where $z \in {\rm I\!R}^n$, $u \in {\rm I\!R}^m$, $y \in {\rm I\!R}^r$ are state, the control and output vectors, respectively; z_o is an initial vector which belongs to the set of admissible initial vectors, U is the set of admissible control functions, p is the vector of unknown parameters in the set of possible values $Q \subset {\rm I\!R}^k$; f and g are (nonlinear) functions representing the dynamic and the observation.

Let us collect some assumptions:

(12.2)

    i) $U = \{u \in L_\infty(0,T;{\rm I\!R}^m) \,|\, u(t) \in \Omega \text{ a.e.}\}$ where Ω is a neighbourhood of Θ;

    ii) $g : {\rm I\!R}^n \times Q \longrightarrow {\rm I\!R}^r$ is continuous;

    iii) $f : [0,T] \times {\rm I\!R}^n \times \Omega \times Q \longrightarrow {\rm I\!R}^n$ is continuous;

    iv) For each $p \in Q$ and each $u \in U$ there exists $\alpha \in L_1(0,T)$ such that

$$|f(t,z_1,u(t);p) - f(t,z_2,u(t);p)| \leq \alpha(t) |z_1 - z_2|$$

for all $t \in [0,T]$, $z_1,z_2 \in {\rm I\!R}^n$.

It is known that under the assumptions above for each $p \in Q$, $u \in U$ and $z_o \in Z_o$ the initial value problem

(12.3) $\qquad z' = f(t,z,u;p)$, $z(0) = z_0$

has a uniquely determined solution z which is absolutely continuous and satisfies the differential equation almost every-
where. Therefore the solution operator

$$L : U \times Z_0 \times Q \ni (u,z_0,p) \longmapsto z \in C(0,T;\mathbb{R}^n), \; z \text{ solves (12.3)} ,$$

and the input-output mapping

$$\mathbb{L} : U \times Z_0 \times Q \ni (u,z_0,p) \longmapsto g(L(u,z_0,p);p) \in C(0,T;\mathbb{R}^r)$$

are well defined.

Let us call $E := U \times Z_0$ the set of <u>admissible experiments.</u>
In designing identification experiments, the first question
that arises is wether or not the unknown parameter p can be
uniquely identified from a proposed experiment $(u,z_0) \in E$.

<u>Definition 12.1</u>

    1)   A parameter $q \in Q$ is said to be <u>identifiable by the</u>
       <u>experiment</u> $(u,z_0) \in E$ if

$$\mathbb{L}(u,z_0,p) \neq \mathbb{L}(u,z_0,q) \text{ for all } p \in Q, \; p \neq q.$$

    2)   A parameter $q \in Q$ is said to be <u>identifiable</u>
       if q is identifiable by some experiment $(u,z_0) \in E$.

From the following simple example it becomes clear that the
experiment (u,z_0) has to be "sufficiently rich" in order to
lead to identifiability.

<u>Example 12.2</u>

    Consider

$$z' = pz + u \;\; , \;\; z(0) \in Z_0 := \{0\}.$$
$$y = z \qquad\quad , \; p \in Q := \mathbb{R}.$$

Then $y(t) = \int_o^t e^{p(t-s)} u(s)\,ds$, $t \geq 0$, and we see that the experiment $u := 0$ doesn't distinguish the parameters. An experiment which identifies every parameter is given by $u \equiv 1$. $*$

<u>Theorem 12.3</u>

Let $(u,z_o) \in E$ and suppose that in addition to the assumptions (12.2) the following assumptions are satisfied:

1) f has infinitely many derivatives with respect to time, the state and the control vector components.

2) g has infinitely many derivatives with respect to the state vector component.

3) The control function is differentiable of arbitrary order.

Then a sufficient condition for the identifiability of the parameter $q \in Q$ by the experiment (u,z_o) is that the set of equations

$$(12.4) \quad \frac{\partial^k}{\partial t^k}\, \mathbb{L}\,(u,z_o,p)\Big|_{t=o} = \frac{\partial^k}{\partial t^k}\, \mathbb{L}\,(u,z_o,q)\Big|_{t=o} \quad , \ k = 0,1,\ldots$$

has only the solution $p = q$.

<u>Proof:</u> The result is an obvious consequence of Definition 12.1.

$\square$

<u>Example 12.4</u>

Consider the model described by

$$z_1' = -p_1 z_1 - p_2(1 - p_3 z_2)z_1 + u \ , \quad z_1(0) = 1$$

$$z_2' = p_2(1 - p_3 z_2)z_1 - p_4 z_2 \quad , \quad z_2(0) = 0$$

$$y = z_1 .$$

Then we obtain with $u \equiv 0$ for $y := \mathbb{L}\,(u,z_o,p)$

$$y(0) = 1, \quad y'(0) = -(p_1 + p_2), \quad y''(0) = (p_1 + p_2)^2 + p_2^2 \, p_3 \, ,$$

$$y'''(0) = -(p_1 + p_2)\{p_2^2 \, p_3 + (p_1 + p_2)^2\} - 2p_2^2 \, p_3 (p_1 + p_2)$$

$$+ \, p_2 \, p_3 \{p_2^2 \, p_3 - p_2 (p_1 + p_2 + p_4)\}.$$

This shows that the following quantities are uniquely determined by the output:

$$p_1 + p_2, \quad p_2^2 \, p_3, \quad p_2 p_3 \{p_2^2 \, p_3 - p_2 (p_1 + p_2 + p_4)\}.$$

To conclude that a parameter $q = (q_1, \ldots, q_4)$ is uniquely identifiable by the chosen experiment at most $y^{(iv)}(0)$ has to be computed. $\qquad *$

As the example above suggests even for relatively simple models the computation by hand of the quantities $\mathbb{L}(u, z_o, p)^{(k)}(0)$ becomes practically impossible. Therefore the approach of computing these quantities by symbolic manipulation on a computer should be used.

12.2 Identification in linear dynamic systems

Now we want to discuss the special case of a linear time-invariant system:

$$z' \; = A(p)z + B(p)u \; , \; t \geq 0,$$

$$(12.5) \qquad z(0) = z_o \in Z_o, \; u \in U, \; p \in Q,$$

$$y \;\;\; = C(p)z \; ;$$

here $A(p) \in \mathbb{R}^{n,n}$, $B(p) \in \mathbb{R}^{n,m}$, $C(p) \in \mathbb{R}^{r,n}$, $p \in Q$.
As it is known from the theory of linear systems the solution of the initial value problem

$$z' \; = A(p)z + B(p)u \; , \; z(0) = z_o$$

is given by

$$z(t) = e^{A(p)t} z_o + \int_o^t e^{A(p)(t-s)} B(p)u(s)ds, \; t \geq 0.$$

Hence the output is given by

$$(12.6) \quad y(t) = C(p)e^{A(p)t}t_o + \int_o^t C(p)e^{A(p)(t-s)}B(p)u(s)ds, \quad t \geq 0.$$

The longest-established method to analyse the identifiability question for linear systems (with homogeneous initial values) consists in transforming the output by taking Laplace transforms. This method was already mentioned in Chapter 1.

As a second method we may use the method of computing the quantities $\mathbb{L}(u,z_o,p)^{(k)}(0)$ as in the nonlinear case. Since we have the output $\mathbb{L}(u,z_o,p)$ given by the formula (12.6) explicitely this method can be used more effective.

Let
$$Y_j(p) := C(p)A(p)^j B(p) \ , \quad p \in Q, \ j = 0,1,\ldots$$

In the following we consider the case of zero initial conditions, i.e.
$$z_o = \{\theta\}.$$

Lemma 12.5

Let $p,q \in Q$. Then the following properties are equivalent:

a) $\mathbb{L}(u,\theta,p) = \mathbb{L}(u,\theta,q)$ for all $u \in U$;

b) $Y_j(p) = Y_j(q)$, $j = 0,1,\ldots$;

c) $Y_j(p) = Y_j(q)$, $j = 0,1,\ldots, 2n - 1.$

Proof:

a) $\Longrightarrow$ b) From a) we have

$$\int_o^t \{C(p)e^{A(p)(t-s)}B(p)-C(q)e^{A(q)(t-s)}B(q)\}u(s)ds = 0, \quad t \in [0,T],$$

for all $u \in U$. Since each $u \in U$ may take values in the neighbourhood Ω of θ we obtain by a simple argument

$$C(p)e^{A(p)t}B(p) = C(q)e^{A(q)t}B(q) \ , \quad t \in [0,T].$$

This implies the identities in b).

The implication b) $\implies$ a) is immediately clear. Therefore it remains to prove the implication c) $\implies$ b). Let $\tilde{p} \in Q$. By the theorem of Caley-Hamilton there exist numbers $\alpha_o(\tilde{p}), \ldots, \alpha_{n-1}(\tilde{p}) \in \mathbb{C}$ with

$$A(\tilde{p})^n = \sum_{i=o}^{n-1} \alpha_i(\tilde{p}) A(\tilde{p})^i.$$

From this follows

$$(1) \qquad Y_k(\tilde{p}) = \sum_{i=o}^{n-1} \alpha_i(\tilde{p})\, Y_{i+k-n}(\tilde{p}) \ , \ k \geq n.$$

Now apply this to p and q and let $\varepsilon_i := \alpha_i(p) - \alpha_i(q)$, $0 \leq i \leq n - 1$. We prove by induction for all $N \geq 2n$ the following two identities:

$$(2) \qquad Y_j(p) = Y_j(q) \ , \ j = 0, \ldots, N-1 \ ;$$

$$(3) \qquad Y_j(p) - Y_j(q) = \sum_{i=o}^{n-1} \varepsilon_i Y_{j-n+i}(p) \ , \ j = n, \ldots, N.$$

The identities (2) and (3) are valid for $N = 2n$ by (1). Let us prove (2),(3) for $N + 1$. We obtain with (1)

$$
\begin{aligned}
Y_N(p) - Y_N(q) &= \sum_{i=o}^{n-1} \varepsilon_i Y_{N-n+i}(p) \\
&= \sum_{i=o}^{n-1} \varepsilon_i \sum_{l=o}^{n-1} \alpha_l(p) Y_{N-n+i-n+1}(p) \\
&= \sum_{l=o}^{n-1} \alpha_l(p) \sum_{i=o}^{n-1} \varepsilon_i Y_{N+i-2n+1}(p) \\
&= \sum_{l=o}^{n-1} \alpha_l(p) \left(Y_{N+1-n}(p) - Y_{N+1-n}(q)\right) = 0
\end{aligned}
$$

and

$$
\begin{aligned}
Y_{N+1}(p) - Y_{N+1}(q) &= \sum_{i=o}^{n-1} \left\{\alpha_i(p) Y_{N+1-n+i}(p) - \alpha_i(q) Y_{N+1-n+i}(q)\right\} \\
&= \sum_{i=o}^{n-1} \varepsilon_i Y_{N+1-n+i}(p).
\end{aligned}
$$

Now, b) follows from (3). $\qquad\qquad\square$

Theorem 12.6

Let $q \in Q$ and $E = \{\theta\} \times U$. Then the following properties are equivalent.

a) q is identifiable.

b) $\forall p \in Q \; \exists j \in \{0,\ldots,2n-1\}(Y_j(p) \neq Y_j(q))$.

Proof: The result is an immediate consequence of Lemma 12.5.

$\square$

Example 12.7

Consider the model

$$z_1' = -(p_1 + p_2)z_1 + p_2 z_2 + u \quad , \; z_1(0) = 0 \; ,$$
$$z_2' = p_2 z_1 - p_3 z_2 \qquad\qquad , \; z_2(0) = 0 \; ,$$
$$y \; = z_1$$

where $Q = \{p = (p_1, p_2, p_3) \in \mathrm{IR}^3 \,|\, p_i \geq 0, \; i = 1,2,3\}$. We obtain

$$Y_0(p) = \; Y_1(p) = -(p_1 + p_2), \; Y_2(p) = (p_1 + p_2)^2 + p_2^2$$
$$Y_3(p) = -(p_1 + p_2)(p_1^2 + 2p_1 p_2 + 3p_2^2) - p_2^2 p_3.$$

From this we conclude that any $q \in Q$ with $q_3 > 0$ is identifiable.

$*$

Remarks 12.8

1) Notice that Th.12.3 may also be applied to the linear model (12.5).

2) The matrix

$$M(p) = \begin{pmatrix} Y_0(p) \\ \vdots \\ Y_{2n-1}(p) \end{pmatrix} \in \mathrm{IR}^{2nr,m}$$

is called the Markov parameter matrix. Clearly, each parameter

$q \in Q$ is identifiable if the map M from the parameter space into the Markov parameters is one to one. This is locally the case if the Jacobian $\frac{\partial M}{\partial p}$ has full rank.

3) If O denotes the so-called observability matrix

$$O(p) = \begin{pmatrix} C(p) \\ \cdot \\ \cdot \\ \cdot \\ C(p)A^{n-1}(p) \end{pmatrix}$$

we have

$$M(p) = \begin{pmatrix} O(p)B(p) \\ C(p)A^{n}(p)B(p) \\ \cdot \\ \cdot \\ \cdot \\ C(p)A^{2n-1}(p)B(p) \end{pmatrix} \qquad . \qquad *$$

12.3 Identification in bilinear structures

Let there be given Hilbert spaces H and $I\!H$ with inner products $(.\,,.)_H$ and $(.\,,.)_{I\!H}$ respectively; moreover, let V and $I\!P$ are Hilbert spaces satisfying

$$V \subset H \subset V^*, \quad I\!P \subset I\!H \subset I\!P^*$$

with dense and continuous imbeddings (see Chapter 5).
Let $Q \subset I\!P$ and let $G : Q \times V \longrightarrow V^*$ be a bilinear mapping. Then the identification problem that we want to consider consists in the following:

(12.7) Given $u \in V$, $y \in V^*$, find $q \in Q$ such that

$$G(q,u) = y.$$

<u>Example 12.9</u>

Consider the boundary value problem

$$\begin{cases} -(qu')' & = f \quad , \text{ in } \Omega := (0,1) \\ -q(0)u'(0) = \beta_o \, , \ u(1) = 0 \end{cases} \quad .$$

The identification of q from the "state" u can be formulated in the above setting by choosing

$$H := I\!H := L_2(0,1), \ V := \{v \in H^1(0,1) \,|\, v(1) = 0\} \, ,$$

$$I\!P := H^1(0,1), \ Q := \{q \in I\!R \,|\, \nu < q(x) < \mu, \ x \in (0,1)\} \text{ with } 0 < \nu < \mu,$$

$$G(q,u)(v) := (qu',v')_H, \ y(v) := \beta_o v(0) + (f,v)_H.$$

Notice that the formal expression

$$q(x) = \frac{-1}{u'(x)} \, \{\beta_o + \int_0^x f(\xi)d\xi\}, \ x \in (0,1),$$

makes clear the difficulties in reconstructing q from the "observation" u and the input f, β_o. A similar problem in one and two dimensions was already mentioned in Chapter 1.

*

Let us define (see Section 5.1)

$$a(q,u,v) := (G(q,u),v)_H \, , \ q \in Q, \ u,v \in V,$$

and consider the following properties:

$$\begin{aligned} &\text{i) } \forall E > 0 \ \ \exists c_1(E) \geq 0 \ \ \forall q \in Q, \|q\|_{I\!P} \leq E \ \forall \, u,v \in V \\ (12.8) \quad &\qquad (\,|a(q,u,v)| \, \leq \, c_1(E)\|u\|_V\|v\|_V) \, . \\ &\text{ii) } \exists c_2 > 0 \ \ \forall q \in Q \ \ \forall u \in V \ \ (a(q,u,u) \geq c_2\|u\|_V^2) \, . \end{aligned}$$

<u>Lemma 12.10</u>

Assume that (12.8) is satisfied. Then the equation

$$G(q,u) = y$$

has for each $q \in Q$ and $y \in V^*$ a unique solution u with

$$\|u\|_V \leq c_2^{-1} \|y\|_{V^*} \quad .$$

<u>Proof:</u> Let $J_V : V^* \to V$ denote the Riesz mapping. Then solving $G(q,u) = y$ is equivalent to solving the equation

$$J_V G(q,u) = J_V y.$$

We shall show that this equation has one and only one solution by showing that the affine mapping

$$(*) \quad V \ni v \longmapsto v - s \cdot (J_V G(q,v) - J_V y) \in V$$

is a contraction if s is small enough. To see this, we observe that with $E := \|q\|_{\mathrm{I\!P}}$

$$\|v - s J_V(q,v)\|^2 \leq (1 - 2sc_2 + s^2 c_1(E)^2)\|v\|^2.$$

Therefore the mapping defined in $(*)$ is a contraction whenever the number s belongs to the interval $(0, c_1(E)^{-2} \, 2c_2)$.

If u is a solution then

$$c_2\|u\|_V^2 \leq a(q,u,u) = (y,u)_H \leq \|y\|_{V^*}\|u\|_V$$

and the given estimate follows. $\qquad\qquad\square$

Thus, if (12.8) is satisfied, the solution operator

$$L : Q \longrightarrow V$$

defined by

$$G(q,Lq) = y \quad , \quad q \in Q,$$

is well defined. Notice that L depends on the given $y \in V^*$ which represents the "experiment" from which we want to identify the parameter.

<u>Definition 12.11</u>

A parameter $\bar{q} \in Q$ is said to be <u>identifiable by the experiment</u> y if and only if $Lq = L\bar{q}$, $q \in Q$, implies $q = \bar{q}$.

In the following we fix the "experiment" y and call a parameter $\bar{q} \in Q$ identifiable if it is identifiable by the given "experiment" y.

Consider the following condition:

> There exists a bilinear mapping $b : V \times V \longrightarrow \mathbb{P}^*$
> such that

$$(12.9) \qquad a(q,u,v) = (q,b(u,v))_{\mathbb{H}}$$

> for all $q \in Q, \quad u,v \in V.$

Theorem 12.12

Let $\bar{q} \in Q$, $\bar{u} := L\bar{q}$, and suppose that (12.8) and (12.9) are satisfied. Then

a) $\bar{q}$ is identifiable if the range of $b(\bar{u},.)$ is dense in $\mathbb{P}^*$.

b) If $\bar{q} \in \text{int } Q$ and $\bar{q}$ is identifiable then the range of $b(\bar{u},.)$ is dense in $\mathbb{P}^*$.

Proof:

a) Let $q \in Q$ with $Lq = \bar{u}$. Then we have for all $v \in V$

$$0 = a(q,Lq,v) - a(\bar{q},\bar{u},v) = (q,b(Lq,v))_{\mathbb{H}} - (\bar{q},b(\bar{u},v))_{\mathbb{H}}$$
$$= (q - \bar{q}, b(\bar{u},v))_{\mathbb{H}}$$

from which the result follows.

b) Assume that the range of $b(\bar{u},.)$ is not dense in $\mathbb{P}^*$. Then there exists $r \in \mathbb{P}$ such that

$$(r,b(\bar{u},v)) \neq 0 \quad \text{for all } v \in V.$$

Then if we set $q := \bar{q} + \alpha r$ we have $q \in Q$ if α is small enough, and we obtain from the identifiability of $\bar{q}$

$$0 = a(r,b(\bar{u},v)) \quad \text{for all } v \in V$$

which is a contradiction.

$\square$

Remark 12.13

Clearly, the property

$$\text{range of } b(\overline{u},.) \text{ is dense in } \mathbb{P}^*$$

has to do with the solvability of the equation

$$b(\overline{u},v) = w \quad (v \in V)$$

for sufficiently many $w \in \mathbb{P}^*$. In concrete situations, this can be ensured by imposing some regularity and descriptive assumptions on $\overline{u}$. From the practical point of view, one should know which assumptions on the "experiment" y these assumptions imply.

Now, we come to the important stability question: Is it possible to estimate $\|q - \overline{q}\|_{\mathbb{H}}$ by a term which depends on $\|Lq - L\overline{q}\|_{H}$?

Let $(H_s)_{s \in \mathbb{R}}$ and $(\mathbb{H}_t)_{t \in \mathbb{R}}$ be Hilbert scales such that

$$H = H_0, \quad V = H_1, \quad \mathbb{H} = \mathbb{H}_0, \quad \mathbb{P} = \mathbb{H}_1.$$

Consider the following condition:

(12.10)   There exists $\alpha > 1$ such that the following properties are satisfied:

     i) $R(L) \subset H_\alpha$.

     ii) $\forall E > 0, \exists c_5(E) \geq 0 \quad \forall q \in Q, |q|_{\mathbb{P}} \leq E$.
       $(\|Lq\|_{H_\alpha} \leq c_5(E))$.

Theorem 12.14

Let $\overline{q} \in Q$, $\overline{u} := L\overline{q}$, and suppose that (12.8),(12.9) and (12.10) are satisfied. Moreover, assume that there exists a mapping $T : Q \longrightarrow V$ such that the following properties hold:

     i) $\exists \beta \in (0,1] \exists c_3 \geq 0 \quad \forall q \in Q (\|Tq\|_V \leq c_3 |q|_{\mathbb{H}_\beta})$.

(12.11)

     ii) $\exists c_4 > 0 \quad \forall q \in Q (c_4 \|q - \overline{q}\|_{\mathbb{H}}^2 \leq (q - \overline{q}, b(\overline{u}, T(q - \overline{q})))_{\mathbb{H}})$.

Then

    a) $\bar{q}$ is identifiable.

    b) $\forall E > 0 \quad \exists c \geq 0 \quad \forall q \in Q, \quad \|q\|_{I\!P} \leq E$

$$\left(\|q - \bar{q}\|_{I\!H} \leq CE^{\frac{\beta}{1+\beta}} \|Lq - \bar{u}\|_H^{\frac{\alpha-1}{\alpha} \cdot \frac{1}{1+\beta}} \right) .$$

__Proof:__   Let $q \in Q$ and $u := Lq$. If $u = \bar{u}$ then we obtain from (12.10)ii) that $q = \bar{q}$ which shows that $\bar{q}$ is identifiable. Therefore part a) is proved. To prove part b) we observe that

$$c_4 \|q - \bar{q}\|_{I\!H}^2 \leq (q - \bar{q}, \, b(\bar{u}, T(q - \bar{q})))_{I\!H}$$

$$= a(q - \bar{q}, \bar{u}, T(q - \bar{q}))$$

$$= - a(q, u - \bar{u}, T(q - \bar{q}))$$

$$\leq c_1(E) \, \|\bar{u} - u\|_V \|T(q - \bar{q})\|_V$$

$$\leq c_1(E) c_3 \|q - \bar{q}\|_{I\!H_\beta} |u - \bar{u}|_V$$

$$\leq c_1(E) c_3 \|q - \bar{q}\|_{I\!H}^{1-\beta} \|q - \bar{q}\|_{I\!P}^{\beta} \cdot$$

$$\cdot \|u - u\|_{H_\alpha}^{\frac{1}{\alpha}} \|u - \bar{u}\|_H^{1-\frac{1}{\alpha}}$$

$$\leq c'(E) E^{\beta} \|q - \bar{q}\|_{I\!H}^{1-\beta} |u - \bar{u}|_H^{1-\frac{1}{\alpha}}$$

where we have used interpolation inequalities.

           □

__Remark 12.15__

    It is important to have an estimate of $\|q - \bar{q}\|_{I\!H}$ in terms of $\|Lq - \bar{u}\|_H$ since then it is possible to consider perturbed data $\bar{u}^\varepsilon$ in H.   *

<u>Example 12.16</u>

Let us consider the problem formulated in Ex.12.9. In this situation we have $b(u,v) = u'v'$ and a parameter $\bar{q} \in Q$ is identifiable if we have for $\bar{u} := L\bar{q}$

$$(*) \qquad \frac{1}{\bar{u}'} \in L_2(0,1)$$

since in this case $H^1(0,1)$ is contained in the range of $b(\bar{u},.)$. If $(*)$ is satisfied then the mapping T is constructed in an obvious manner:

$$Tq(x) := \int_1^x \bar{u}'(\xi)q(\xi)d\xi \ , \ x \in [0,1].$$

One can show that we can choose $\alpha = 2$ and $\beta = 1$ if $y \in L_2(0,1)$.

Finally, let us give a short overview of the methods applicable to solve the identification problem (12.7). Let $\bar{u} = L\bar{q}, \bar{q} \in Q$. We distinguish three groups of methods for the reconstruction of $\bar{q}$:

- Equation error methods;
- Output error methods;
- Methods based on transformations.

The <u>equation error method</u> consists in minimizing the error criterion

$$\|G(q,\bar{u}) - y\|_{V^*}$$

over the set of admissible parameters Q. This minimization can be carried out in a continuous or a discrete version (Ritz method). Clearly, difficulties arise if we have at hand only a noisy measurement $\bar{u}^\varepsilon \in H$ for $\bar{u}$. To avoid the obvious problems one can apply a smoothing or filtering procedure to the data $\bar{u}^\varepsilon$. The discretization of the problem has in some sense the same effect.

In the <u>output (or response) error method</u> the parameters are chosen so that they minimize the error criterion

$$\|Lq - \bar{u}\|_H$$

over the set of admissible parameters Q. Again, one can do
this in a continuous and a discrete form. As one sees immediately, formally there is no problem to use this method also if
we have only noisy data $\bar{u}^{\varepsilon}$ at hand. Of course, the difficulty
in the noisy case is the establishment of conditions for the
existence of minimizers and convergence of algorithms.
If we assume that we have the exact data $\bar{u}$, the minimization
of $\|Lq - \bar{u}\|_H$ may also be carried out by solving the equations

$$G(q,u) = y$$
$$u = \bar{u}$$

or alternatively

$$(12.12) \qquad \begin{aligned} G(q,u) &= y \\ -b(u,u-\bar{u}) &= 0. \end{aligned}$$

If one applies the continuous successive approximation method
one obtains the following system

$$(12.13) \qquad \begin{aligned} \dot{u}(t) + G(q(t),u(t)) &= y &, u(0) &= \bar{u} \\ & &,t > 0 \\ \dot{q}(t) - b(u(t),u(t)-\bar{u}(t)) &= 0 &, q(0) &= q_0 \end{aligned}$$

where q_0 is an initial guess for the true parameter $\bar{q}$ and the
dot denotes derivation with respect to the continuous "iteration" parameter t. The aim is to prove that

$$\lim_{t\to\infty} u(t) = \bar{u} , \quad \lim_{t\to\infty} q(t) = \bar{q}.$$

The analysis of the system (12.13) is difficult since it is a
nonlinear system. An alternative which admits a linear theory
is constructed as follows:

$$\begin{aligned} &\text{Choose a "good" parameter } q^* \in Q \text{ and} \\ &\text{consider the equations} \end{aligned}$$

$$(12.14) \qquad \begin{aligned} G(q^*,u) + G(q,\bar{u}) - G(q^*,\bar{u}) &= y \\ - b(\bar{u},u - \bar{u}) &= 0. \end{aligned}$$

Clearly, $(\bar{u},\bar{q})$ solves these equations. If we now apply the
continuous successive approximation method one obtains the

following system:

$$\dot{u}(t) + G(q^*, u(t)) + G(q(t), \bar{u}) = y + G(q^*, \bar{u}) \quad , \quad u(0) = u$$

(12.15)
$$\dot{q}(t) - b(\bar{u}, u(t) - \bar{u}) = 0 \quad , \quad q(0) = q_0 \qquad , t > 0$$

This system is easier to analyse since the "error" quantities
$w := u - \bar{u}$, $r := a - \bar{a}$ are governed by the linear system

$$\dot{w} + G(q^*, w) + G(r, \bar{u}) = 0 \quad , \quad w(0) = 0$$

(12.16)
$$\dot{r} - b(\bar{u}, w) = 0 \quad , \quad r(0) = r_0$$

Example 12.17

Consider again the problem (see Ex.12.9 and Ex.12.16)

$$\begin{cases} -(qu')' = f & , \text{ in } \Omega := (0,1), \\ -q(0)u'(0) = \beta_0 \,, \; u(1) = 0 \end{cases} \qquad .$$

Identification by the method (12.13): Solve

$$\dot{u} - (qu')' = f$$
$$\dot{q} - u'(u' - \bar{u}') = 0 \qquad , \text{ in } (0,\infty) \times (0,1),$$

$$u(0,x) = \bar{u}(x), \; q(0,x) = q_0(x) \quad , \quad x \in (0,1),$$

$$-q(t,0)u'(t,0) = \beta_0, \; u(t,1) = 0, \; t \geq 0.$$

Identification by the method (12.15): Solve

$$\dot{u} - (q^*u')' - (q\bar{u}')' = f - (q^*\bar{u}')'$$
$$\dot{q} - \bar{u}'(u' - \bar{u}') = 0 \qquad , \text{ in } (0,\infty) \times (0,1) \,,$$

$$u(0,x) = \bar{u}(x), q(0,x) = q_0(x) \qquad , \; x \in (0,1),$$

$$-q^*(t,0)u'(t,0) + (q(t,0) - q^*(t,0))\bar{u}'(t,0) = \beta_0, \; u(t,1) = 0, \; t \geq 0.$$

$$*$$

In the next section we come back to a similar system in a special case.

The main __transformations__ on which methods for identifi-cation may be based are the Fourier transform, the Laplace transform and the Mellin transform. In the image domain

(Fourier-, Laplace-, Mellin-domain) one can apply the equation error and the output error method.

12.4 Adaptive identification

In this section we consider a time-invariant linear system of the form

$$(12.17) \qquad z' = Az + f(t) \; , \; t \geq 0,$$

where $A \in \mathbb{R}^{n,n}$, $z \in \mathbb{R}^n$, $f : [0,\infty) \longrightarrow \mathbb{R}^n$ and t is the time. The aim of the identification procedure that we want to describe is to determine the elements of A from the observation of the states $z(t)$, $t \geq 0$. As a model for the process (12.17) we consider

$$u'(t) = Cu(t) + (M(t)-C)z(t) + f(t), \; t \geq 0,$$

where $C \in \mathbb{R}^{n,n}$ is a known stability matrix $(\max\{\text{Re}(\lambda) \,|\, \lambda$ eigenvalue of $A \} < 0)$ and $M(t) \in \mathbb{R}^{n,n}$ is a matrix of adjustable parameters. It is desired to determine adaptive laws for adjusting the time derivative $M'(t)$, $t \geq 0$, so that

$$\lim_{t\to\infty} u(t) - z(t) = 0, \quad \lim_{t\to\infty} M(t) = A.$$

Let $w(t) := u(t) - z(t)$, $R(t) := M(t) - A$, $t \geq 0$, and let us consider the following adaptation law:

$$M'(t) = F(t,u(t),z(t)) \; , \; t \geq 0.$$

Then, since $R'(t) = M'(t)$, $t \geq 0$, we are led to the following system for the error quantities w and R:

$$(12.18) \quad \begin{aligned} w' &= Cw + Rz(t) \\ R' &= F(t,w + z(t),z(t)) \end{aligned} \; , \; t \geq 0.$$

Now, we want to determine the adaption law F in such a way that the zero vector in $\mathbb{R}^n \times \mathbb{R}^{n,n}$ is a globally asymptotically stable equilibrium point. Notice that the system (12.18) is a time-varying system. To design the adaptation law F we use the approach of Lyapunov functions. As a Lyapunov function candidate we define

$$V(w,R) := \tfrac{1}{2} w^t \Gamma w + \tfrac{1}{2} tr(R^t R)$$

where $\Gamma \in \mathbb{R}^{n,n}$ is a symmetric positive definite matrix. Let

$$\dot{V} := \frac{\partial V}{\partial w} w' + \frac{\partial V}{\partial R} R' = \frac{\partial V}{\partial w}(Cw + Rz) + \frac{\partial V}{\partial R} F.$$

We want to show that $\dot{V}$ is negative semidefinite. A simple calculation gives in a short notation

$$\dot{V} = \tfrac{1}{2} w^t (C^t \Gamma + \Gamma C^t) w + tr(R^t \Gamma w z^t) + tr(R^t F).$$

The first term on the rigth-hand side will be negative semidefinite if

$$N := -C^t \Gamma - \Gamma C^t$$

is a positive definite matrix, and the second and third term will be identically zero if one choses the adaption law

$$M' = -\Gamma w z^t.$$

Thus, we arrive at the following system for (w,R):

$$(12.19) \qquad \begin{cases} w' = Cw + Rz(t) \\ R' = -\Gamma w z(t)^t \end{cases} , \quad t \geq 0.$$

For the original quantities (u,M) the following system results:

$$(12.20) \qquad \begin{cases} u' = Cu + (M-C)z(t) + f(t) &, u(0) = z(0) \\ M' = \Gamma(u - z(t))z(t)^t &, M(0) = A_o \end{cases} , \quad t > 0 ,$$

where A_o is an initial guess for the true parameter A.

Let us collect assumptions that we need for a first step in the analysis:

$$(12.21)$$

    i) C stability matrix, Γ positive definite,

$$N := -C^t \Gamma - \Gamma C^t \text{ positive definite.}$$

    ii) $f \in C((0,\infty);\mathbb{R}^n) \cap L_\infty((0,\infty);\mathbb{R}^n)$,

$$z \in C^1((0,\infty);\mathbb{R}^n) \cap L_\infty((0,\infty);\mathbb{R}^n).$$

Under these assumptions the system (12.19) has a local solution (w,R). For such a solution we obtain

$$V(w(t),R(t)) - V(w(0),R(0)) = -\frac{1}{2} \int_0^t w(s)^t N w(s)\, ds$$

and therefore

$$\frac{1}{2} w(t)^t \Gamma w(t) + \frac{1}{2} \operatorname{tr}(R(t)^t R(t)) + \frac{1}{2} \int_0^t w(s)^t N w(s)\, ds$$

(12.22)

$$= \frac{1}{2} w(0)^t \Gamma w(0) + \frac{1}{2} \operatorname{tr}(R(0)^t R(0))$$

<u>Lemma 12.18</u>

Let the assumption (12.21) be satisfied. Then the system (12.19) has a unique solution $(w,R) : [0,\infty) \longrightarrow \mathbb{R}^n \times \mathbb{R}^{n,n}$ which has the following properties

i) $\displaystyle\sup_{t \geq 0} \{|w(t)|^2 + \operatorname{tr}(R^t(t)R(t))\} + \int_0^\infty |w(s)|^2\, ds < \infty$.

ii) $(w,R) \in C^1((0,\infty); \mathbb{R}^n \times \mathbb{R}^{n,n})$.

iii) $w \in L_\infty((0,\infty); \mathbb{R}^n) \cap L_2((0,\infty); \mathbb{R}^n)$.

iv) $R \in L_\infty((0,\infty); \mathbb{R}^{n,n}), R' \in L_\infty((0,\infty); \mathbb{R}^{n,n})$.

v) $\displaystyle\lim_{t \to \infty} \frac{1}{2} w(t)^t \Gamma w(t) + \frac{1}{2} \operatorname{tr}(R(t)^t R(t))$ exists.

<u>Proof:</u> The inequality (12.22) shows that the local solution (w,R) can be continued uniquely for all times $t \geq 0$; the property i) is then a consequence of (12.22). The properties ii), iii) and iv) follow from (12.22) and the system (12.19) for (w,R). The last assertion is a consequence of the fact that the Lyapunov function V along the trajectory (w,R) is nonincreasing.

$\square$

Clearly, since a unique solution (w,R) exists the system (12.20) has also a unique solution (u,M).

<u>Theorem 12.19</u>

Suppose that (12.21) is satisfied and that (w,R) is the solution of (12.19). Then

$$\lim_{t \to \infty} w(t) = 0.$$

<u>Proof:</u> From (12.19) we obtain

$$\frac{1}{2} \frac{d}{dt} w(t)^t \Gamma w(t) = -\frac{1}{2} w(t)^t N w(t) + w(t)^t \Gamma^t R(t) z(t)$$

and therefore by Lemma 12.18

$$|w(t_2)^t \Gamma w(t_2) - w(t_1)^t \Gamma w(t_1)|$$

$$\leq c_1 \int_{t_1}^{t_2} |w(s)|^2 ds + c_2 (t_2 - t_1)^{1/2} \left(\int_{t_1}^{t_2} |w(s)|^2 ds \right)^{1/2}$$

for $t_2 \geq t_1$. This gives the following assertion:

$$(1) \quad \mathop{\forall}_{\rho > 0} \ \mathop{\forall}_{\varepsilon > 0} \ \mathop{\exists}_{t_o \geq 0} \ \mathop{\forall}_{t_1, t_2 \geq t_o} \quad (|t_1 - t_o| < \rho$$

$$\Rightarrow |w(t_2)^t \Gamma w(t_2) - w(t_1)^t \Gamma w(t_1)| < \varepsilon).$$

Assume by contradiction that $(w(t)^t \Gamma w(t))_{t \geq 0}$ doesn't converge to zero as t goes to infinity. Then there exist $\delta > 0$ and a sequence $(t_n)_{n \in \mathbb{N}}$ with

$$(2) \quad \lim_n t_n = \infty; \ t_{n+1} - t_n \geq 2, \ w(t_n)^t \Gamma w(t_n) \geq \delta, \ n \in \mathbb{N}.$$

Now apply (1) with $\rho = 1$, $\varepsilon = \frac{\delta}{2}$. Then there must exist $t_o \geq 0$ and $n_o \in \mathbb{N}$ such that for all $t \in (t_n - 1, t_n + 1)$ and $n \geq n_o$

$$w(t)^t \Gamma w(t) \geq \frac{\delta}{2}.$$

This implies

$$\int_0^\infty |w(s)|^2 ds \geq c_3 \int_0^\infty w(s)^t \Gamma w(s) ds$$

$$\geq c_3 \sum_{n=n_o}^\infty \int_{t_n - 1}^{t_n + 1} w(s)^t \Gamma w(s) ds$$

$$\geq c_3 \sum_{n=n_o}^\infty 2 \cdot \frac{\delta}{2} = \infty$$

which is a contradiction by Lemma 12.18. $\square$

<u>Corollary 12.20</u>

Suppose that (12.21) is satisfied and that (w,R) is the solution of (12.19). Then

i) $\lim\limits_{t\to\infty} \mathrm{tr}(R(t)^t R(t))$ exists, $\lim\limits_{t\to\infty} R'(t) = \Theta$.

ii) $R' \in L_2((0,\infty);\mathbb{R}^{n,n})$.

<u>Proof:</u> Part i) follows from Lemma 12.18 and the fact that $\lim\limits_{t\to\infty} w(t) = \Theta$. Part ii) is true since $w \in L_2((0,\infty);\mathbb{R}^n)$.
$\square$

The convergence of the parameters $M(t), t \geq 0$, requires one to test under what conditions $\lim\limits_{t\to\infty} w(t) = \Theta$ implies $\lim\limits_{t\to\infty} R(t) = \Theta$. This question is certainly connected with the question whether the experiment given by f and $z_0 := z(0)$ distinguishes the parameters by the output. In the following theorem we give a condition which is sufficient for $\lim\limits_{t\to\infty} R(t) = \Theta$. A drawback of this condition is that it is formulated with the state z not with the experiment quantities (f,z_0) itself.

<u>Theorem 12.21</u>

Let the assumption (12.21) be satisfied and suppose that there exist positive numbers ε,δ,τ and a sequence $(t_k)_{k\in\mathbb{N}}$ with $\lim\limits_{k} t_k = \infty$ such that

$$(12.23) \quad \forall_{V\in\mathbb{R}^{n,n},\,|V|=1} \quad \forall_{k\in\mathbb{N}} \quad \exists_{t_k^*\in[t_k,t_k+\tau]} \quad \left(\left|\int_{t_k^*}^{t_n^*+\delta} Vz(s)\,ds\right| \geq \varepsilon\right).$$

Then $\lim\limits_{t\to\infty} R(t) = \Theta$.

<u>Proof:</u> By Cor.12.20 there exist $c := \lim\limits_{t\to\infty}|R(t)|$. Assume by contradiction $c > 0$.

Then there exist $k_o \in \mathbb{N}$ and $\alpha > 0$ such that $|R(t_k)| \geq \alpha$ for all $k \geq k_o$. Without loss of generality we may assume $k_o = 1$. Let $\tilde{t}, t \geq 0$. Then

$$w(\tilde{t}) - w(t) = \int_t^{\tilde{t}} Cw(s)ds + \int_t^{\tilde{t}} R(s)z(s)ds$$

and

$$\left| \int_t^{\tilde{t}} R(s)z(s)ds \right| \leq |w(\tilde{t}) - w(t)| + \left| \int_t^{\tilde{t}} Cw(s)ds \right|$$

$$\leq |w(\tilde{t})| + |w(t)| + c_1 |\tilde{t}-t|^{1/2}$$

$$\left| \int_t^{\tilde{t}} |w(s)|^2 ds \right|^{1/2}.$$

In a similar manner we obtain

$$|R(\tilde{t}) - R(t)| \leq c_2 |\tilde{t}-t|^{1/2} \left| \int_t^{\tilde{t}} |w(s)|^2 ds \right|^{1/2}.$$

Let $(t'_k)_{k \in \mathbb{N}}$ be a sequence with $t'_k \in [t_k, t_k + \tau]$, $k \in \mathbb{N}$. Then

$$(*) \qquad \lim_k \int_{t'_k}^{t'_k+\delta} R(s)z(s)ds = \Theta, \quad \lim_k |R(t_k) - R(t'_k)| = 0,$$

$$\lim_k \int_{t'_k}^{t'_k+\delta} |R(t_k) - R(s)|ds = 0,$$

and

$$\left| \left| \int_{t'_k}^{t'_k+\delta} R(t_k)z(s)ds \right| - \left| \int_{t'_k}^{t'_k+\delta} R(s)z(s)ds \right| \right|$$

$$\leq \int_{t'_k}^{t'_k+\delta} |R(t_k) - R(s)||z(s)|ds$$

$$\leq c_2 \int_{t'_k}^{t'_k+\delta} |R(t_k) - R(s)|ds.$$

Therefore by $(*)$

$$\lim_k \left| \int_{t'_k}^{t'_k+\delta} R(t_k)z(s)ds \right|$$

$$= \lim_k |R(t_k)| \left| \int_{t'_k}^{t'_k+\delta} R(t_k)|R(t_k)|^{-1}z(s)ds \right| = 0.$$

But chosing $V_k := |R(t_k)|^{-1} R(t_k)$ and $t_k^* \in [t_k, t_k + \tau]$ according to assumption (12.23), $k \in \mathbb{N}$, we arrive at a contradiction since

$$|R(t_k)| \, | \int_{t_k^*}^{t_k^* + \delta} R(t_k) |R(t_k)|^{-1} z(s) ds | \geq \alpha\varepsilon > 0, \quad k \in \mathbb{N}. \qquad \square$$

Now the question is which experiments (f, z_o) lead to a state z that has the property formulated in (12.23).

Illustration 12.22

Let us consider the adaptive identification procedure in the following situation:

$$n = 2, \quad A = \begin{pmatrix} -2 & -1 \\ -1 & -4 \end{pmatrix}, \quad z_i(t) = d_i \sin e_i \pi t, \quad i = 1, 2.$$

$$A_o = \begin{pmatrix} 5 & 5 \\ 5 & 5 \end{pmatrix} \quad \text{(initial guess)} .$$

a) We choose $d_1 = d_2 = 1$, $e_1 = 1$, $e_2 = 2$,

$$C = \begin{pmatrix} -1 & 0 \\ 0 & -1 \end{pmatrix}, \quad \Gamma = \begin{pmatrix} 10 & 0 \\ 0 & 10 \end{pmatrix}$$

and obtain the following results:

t	a_{11}	a_{21}	a_{12}	a_{22}	
10.	-1.7972	-0.8228	-0.0305	-2.5496	
20.	-1.9658	-0.9584	-0.7835	-3.6905	
25.	-2.0145	-1.0214	-0.8974	-3.8578	Table 4

b) We choose $d_1 = d_2 = 1$, $e_1 = 1$, $e_2 = 2$,

$$C = \begin{pmatrix} -10 & 0 \\ 0 & -10 \end{pmatrix}, \quad \Gamma = \begin{pmatrix} 50 & 0 \\ 0 & 50 \end{pmatrix}$$

and obtain

t	a_{11}	a_{21}	a_{12}	a_{22}
2.	-1.9291	-0.9440	-0.9975	-3.9907
3.	-2.0076	-1.0059	-1.0014	-4.0014
4.	-1.9992	-0.9993	-1.0001	-4.0001

Table 5

c) We choose $d_1 = 1$, $d_2 = 1$, $e_1 = 1$, $e_2 = 0$,

$$C = \begin{pmatrix} -10 & 0 \\ 0 & -10 \end{pmatrix} \quad , \quad \Gamma = \begin{pmatrix} 50 & 0 \\ 0 & 50 \end{pmatrix}$$

and obtain

t	a_{11}	a_{21}	a_{12}	a_{22}
1.	-2.1138	-1.0975	5.0	5.0
2.	-1.9981	-0.9984	5.0	5.0
3.	-2.0001	-1.0001	5.0	5.0
4.	-2.0	-1.0	5.0	5.0

Table 6

Notice that in c) the richness condition (12.23) is not satisfied.

Bibliographical comments

A complete treatment of compartmental modelling and tracer kinetics is given in ANDERSON [1]; see also GODFREY, DI STEFANO [38]. Th. 12.6 is taken from THOWSEN [98]. The presentation of Section 12.3 follows BAUMEISTER [8]. Equation error methods in identification problems of bilinear structure are discussed in CORAY [22] and SCHABACK [93]. The description of a transformation method can be found in PROVENCHER, VOGEL [84]. The results in Section 12.4 are taken from BAUMEISTER, SCONDO [10]. The problem of identifiability is discussed also in JAQUEZ, GREIF [59] and HADAEGH, BEKEY [49].

REFERENCES

[1] ANDERSON,D.H.:Compartmental modelling and tracer
    kinetics. Lecture Notes Biomathematics,Vol.5o,
    Springer-Verlag, 1983.
[2] ANDERSSEN,R.S.,BLOOMFIELD,P.:Numerical differentiation
    procedures for non-exact data. Numer.Math.22, 157-182,
    1974.
[3] ANG,D.D.: Stabilized approximate solutions of the inverse
    time problem for a parabolic evolution equation. J.Math.
    Anal.Appl. 111 , 148-155,1985.
[4] ANGER,G.(ed.):Inverse and improperly posed problems in
    differential equations. Proc.Conf.Halle, Akademie-Verlag,
    Berlin, 1979.
[5] AREF'EVA,M.V.:Asymptotic estimates for the accuracy of
    optimal solutions of equations of the convolution type.
    Zh.vychisl.Mat.mat.Fiz.14, 838-851,1974.
[6] AREF'EVA,M.V.:Asymptotic optimal error estimates for
    convolution type integral equations of the first kind.
    Zh.vychisl.Mat.mat.Fiz.15, 1310-1317,1975.
[7] BARAKAT,R.,NEWSAM,G.: Algorithms for reconstruction of
    partially known band limited Fourier transform pairs from
    noisy data. I.The prototypical linear problem. J.Integral
    Equations 9,49-76, 1985.
[8] BAUMEISTER,J.: Zur Parameteridentifikation in bilinearen
    Systemen. Preprint, Universität Frankfurt(M),1984.
[9] BAUMEISTER,J.: On the adaptive solution of inconsistent
    systems of linear equations. In:Proc.of the third seminar
    on "Model Optimization and Exploration Geophysics, Berlin,
    1985, to appear.
[10] BAUMEISTER,J.,SCONDO,W.: Adaptive methods for parameter
     identification. In: Methoden und Verfahren der Mathema-
     tischen Physik, Oberwolfach, 1985.
[11] BEN-ISRAEL,A.,GREVILLE,T.N.E.: Generalized matrix inverses:
     Theory and applications. J.Wiley,New York, 1974.
[12] BERTERO,M.,DE MOL,C.,VIANO,G.A.: The stability of inverse
     problems. In: Inverse Scattering Problems in Optics,
     Baltes (ed.), Springer-Verlag, New York, 1980.
[13] BRYNIELSON,L.: On Fredholm integral equations of the first
     kind with convex constraints. SIAM J.Math.Anal.5,955-962,
     1974.
[14] BUBE,K.P.,BURRIDGE,R.: The one-dimensional inverse prob-
     lem of reflection seismology. SIAM Review 25,497-559,
     1983.
[15] BUTLER,J.P.,REEDS,J.A.,DAWSON,S.V.: Estimating solutions
     of first kind integral equations with non-negative con-
     straints and optimal smoothing. SIAM J.Numer.Anal. 18,
     381-397, 1981.
[16] BUZBEE,B.L.,CARASSO,A.: On the numerical computation of
     parabolic problems for preceding times. Math.of Comp.
     27, 237-266,1973.
[17] CAMPELL,S.L., MEYER,C.D.,Jr.: Generalized inverses of
     linear transformations. Pitman,London,1979.

[18] CARASSO,A.,SANDERSON,J.G.,HYMAN,J.M.: Digital removal of random media image degradations by solving the diffusion equation backwards in time. SIAM J.Numer.Anal.15, 344-367, 1978.

[19] CHANDAN,K.,SABATIER,P.:Inverse problems in quantum scattering theory. Springer-Verlag,New York, 1977.

[20] COLLI FRANZONE,P.,GUERRI,L.,MAGENES,E.: Oblique double layer potentials for the direct and inverse problems of electrocardiology. Math.Biosci.68,23-55, 1984.

[21] COLTON,D.,KRESS,R.: Integral equation method in scattering theory. Wiley-Interscience, New York, 1983.

[22] CORAY,C.S.: Use of spline subspaces in minimum norm differential approximation. J.Math.Anal.Appl.64, 159-165, 1978.

[23] CRAVEN,P.,WHABA,G.: Smoothing noisy data with spline functions. Numer.Math.31,377-403, 1979.

[24] CULLUM,J.: Numerical differentiation and regularization. SIAM J.Numer.Anal.8, 254-265, 1971.

[25] DAVIES,A.R.: On the maximum likelihood regularization of Fredholm convolution equations of the first kind. In: Treatment of integral equations by numerical methods, Backer,Miller (eds.), Academic Press, 95-105, 1982.

[26] DEUFLHARD,P.,SAUTTER,W.: On rank-deficient pseudoinverses. Lin.Algebra and its Applications,29, 91-111, 1980.

[27] DEUFLHARD,P.,HAIRER,E.(eds.):Numerical Treatment of Inverse Problems in Differential and Integral Equations. Springer-Verlag, PSC Vol.2, 1983.

[28] DOETSCH,G.: Zerlegung einer Funktion in Gaußsche Fehlerkurven und zeitliche Zurückverfolgung eines Temperaturzustandes. Math.Zeitschrift 42, 263-286, 1937.

[29] ELDÉN,L.: Perturbation theory for the least squares problem with linear equality constraints. SIAM J.Numer.Anal. 17, 338-350, 1980.

[30] ELDÉN,L.: Time discretization in the backward solution of parabolic equations I. Math. Comp. 39, 53-84, 1982.

[31] ELDÉN,L.: A note on the computation of the generalized cross-validation function for ill-conditioned least squares problems. BIT,24, 467-472, 1984.

[32] ELLIOTT,D.F.,RAO,K.R.: Fast Transforms. Algorithms, Analyses, Applications. Academic Press, New York, 1982.

[33] EMELIN,I.V.,KRASNOSEL'SKII,M.A.: A stopping rule in iteration procedures for solving ill-posed problems. Autom. Remote Control 39, 1783-1787, 1978.

[34] ENGL,H.W.: Necessary and sufficient conditions for convergence of regularization methods for solving linear operator equations of the first kind. Numer.Funct.Anal. & Optimiz.3, 201-222, 1981.

[35] ENGL,H.W.,NEUBAUER,A.: Optimal discrepancy principles for the Tikhonov regularization of integral equations. In: Constructive methods for the practical treatment of integral equation.Hämmerlin,Hoffmann(eds.) Birkhäuser, 1985.

[36] FRANKLIN,J.N.: On Tikhonov's method for ill-posed problems. Math.of Comp.28,889-907, 1974.

[37] GABUSHIN,V.N.: Optimal methods of computing the values
of the operator Ux if x is given with an error. Proc.
Steklov.Inst.145(1),67-83, 1981.
[38] GODFREY,K.R.,DI STEFANO,K.: Identifiability of model
parameters. In: IFAC Identification and System Parameter
Estimation, York, 1985.
[39] GOLUB,G.H.,PEREYRA,V.: The differentiation of pseudo-
inverses and non-linear least squares problems whose
variables separate. SIAM J.Numer.Anal.10, 413-432, 1973.
[40] GOLUB,G.H.,HEATH,M.,WAHBA,G.: Generalized cross valida-
tion as a method for choosing a good ridge parameter.
Technometrics 21,215-223, 1979.
[41] GOLUB,G.H.,VAN LOAN,C.F.: Matrix computations. The John
Hopkins University Press, Baltimore, 1983.
[42] GRENACHER,F.: Über die Konvergenz bei Regularisierungsverfahren für nicht sachgemäß gestellte Anfangswertpro-
bleme. Dissertation, Freiburg, 1976.
[43] GROETSCH,C.W.: Generalized inverses of linear operators:
Representation and approximation. Dekker,New York, 1977.
[44] GROETSCH,C.W.: Elements of applicable functional analysis. Marcel Dekker,New York, 1980.
[45] GROETSCH,C.W.: On a class of regularization methods.
Boll.Un.Math.Ital. 17-B, 1411-1419, 1980.
[46] GROETSCH,C.W.: Comments on Morozov's discrepancy principle.
In: Improperly posed problems and their numerical treatment. Hämmerlin,G.,Hoffmann,K.-H.(eds.)Birkhäuser,Basel,1983.
[47] GROETSCH,C.W.: The theory of Tikhonov regularization for
Fredholm equations of the first kind. Pitman, Boston,
1984.
[48] GRÜNBAUM,F.A.: A study of Fourier space methods for
"limited angle" image reconstruction. Numer.Funct.Anal.
and Optimiz. 2, 31-42, 1980.
[49] HADAEGH,F.Y.,BEKEY,G.A.: Near-identifiability of dynam-
ical systems. Math.Biosci.77, 325-340. 1985.
[50] HAMAKER,C.,SOLMON,D.C.: The angles between the null spaces
of X rays. J.Math.Anal.Appl. 62, 1-23, 1978.
[51] HÄMMERLIN,G.,HOFFMANN,K.-H.(eds.): Improperly posed problems and their numerical treatment. Proc.Conf.Oberwol-
fach, Birkhäuser,Basel,1983.
[52] HERBER,M.: Das konjugierte Gradientenverfahren im Hil-
bertraum mit Anwendungen auf Integralgleichungen 1.Art.
Diplomarbeit, Universität Frankfurt(M), 1984.
[53] HERMAN,G.T.,NATTERER,F.(eds.):Mathematical aspects of computerized tomography. Springer-Verlag,New York, 1981.
[54] HÖHN,W.: Finite elements for parabolic equations backwards in time. Numer.Math.40, 207-227, 1982.
[55] HOFMANN,B.: Über Quelldarstellungen bei einigen linearen
Regularisierungsverfahren. Beitr.z.Numer.Math.7, 75-81,
1979.
[56] HOFMANN,B.: Regularization for applied inverse and ill-
posed problems. Teubner-Verlag,Leipzig,1986.
[57] HSIAO,G.C.,WENDLAND,W.: The Aubin-Nitsche lemma for integral equations. J.Integral Equations 3, 299-315, 1981.
[58] HUNT,R.B.: The inverse problem of radiography. Math.
Biosci. 8, 161-179, 1970.

[59] JACQUEZ,J.A.,GREIF,P.: Numerical parameter identifiability
and estimability: Integrating identifiability, estimability,
and optimal sampling design. Math.Biosci.77, 201-227,1985.

[60] JOHN,F.: Continuous dependence on data for solutions of
partial differential equations with a prescribed bound.
Comm.on Pure and Appl.Math.13, 551-585, 1960.

[61] KACZMARZ,S.: Angenäherte Auflösung von Systemen linearer
Gleichungen. Bull.International Acad.Polon.Sci.35, 355-
357, 1937.

[62] KAMMERER,W.J.,NASHED,M.Z.: Iterative methods for best approximation solutions of linear integral equations of the
first and second kinds. J.Math.Anal.Appl. 40, 547-573,1972.

[63] KNABNER,P.: Fragen der Rekonstruktion und der Steuerung bei
Stefan Problemen und ihre Behandlung über lineare Ersatz-
aufgaben. Dissertation, Universität Augsburg, 1983.

[64] KREIN,S.G.,PETUNIN,Ju.I.: Interpolation of linear Operators.
Amer.Math.Soc.,Providence, 1982.

[65] LITTLE,G.,RAEDE,J.B.:Eigenvalues of analytic kernels.
SIAM J.Math.Anal.15, 133-136, 1984.

[66] LOCKER,J.,PRENTER,P.M.: Regularization with differential
operators,I: General theory. J.Math.Anal.Appl.74,504-529,
1980.

[67] LÖTSTEDT,P.: Solving the minimal least squares problem
subject to bounds on the variables. BIT 24,206-224, 1984.

[68] MARTINEZ,J.M.: Solution of nonlinear systems of equations
by an optimal projection method. Computing 37,59-70, 1986.

[69] MELKMAN,A.A.,MICCHELLI,C.A.: Optimal estimation of linear
operators in Hilbert spaces from inaccurate data. SIAM J.
Numer.Anal.16, 87-105, 1979.

[70] MICCHELLI,C.A.: On an optimal method for the numerical
differentiation of smooth functions. J.Approximation
Theory 18, 189-204, 1976.

[71] MICCHELLI,C.A.: Orthogonal projections are optimal algorithms. J.Approximation Theory 40, 101-110, 1984.

[72] MILANESE,M.,TEMPO,R.: Optimal algorithms theory for robust
estimation and prediction. IEEE Trans.on Automatic Control
Vol.AC-30, 730-738, 1985.

[73] MILLER,K.: Least squares methods for ill-posed problems
with a prescribed bound. SIAM J.Math.Anal.1,52-74, 1970.

[74] MOROZOV,V.A.: Methods for solving incorrectly posed problems. Springer-Verlag, New York, 1984.

[75] NASHED,Z.: On moment-discretization and least squares
solutions of linear integral equations of the first kind.
J.Math.Anal.Appl.53, 359-366, 1976.

[76] NATTERER,F.: The finite element method for ill-posed problems. R.A.I.R.O.Analyse numerique 11,271-278, 1977.

[77] NATTERER,F.: Regularisierung schlecht gestellter Probleme
durch Projektionsverfahren. Numer.Math.28,329-341, 1977.

[78] NATTERER,F.: Discretizing ill-posed problems. Publ.dell'
instituto di analisi globale e applicazioni Serie"Problem
non ben posti ed inversi". Florenz, 1983.

[79] NATTERER,F.: Error bounds for Tikhonov regularization in
Hilbert scales. Applic.Analysis 18, 25-37, 1984.

[80] NEUMANN,J.von: The geometry of orthogonal spaces, II.
Princeton University Press, 1950.

[81] NEUMANN-DENZAU,G.,BEHRENDS,J.: Inversion of seismic data
     using tomographical reconstruction techniques for inves-
     tigations of laterally inhomogeneous media. Geophys. J.
     R.Astr.Soc. 305-315, 1984.
[82] PINKUS,A.:n-widths in approximation theory. Springer-Ver-
     lag, New York, 1985.
[83] PORTER,R.P.,DEVANEY,A.J.: Holography and the inverse
     source problem.J.Opt.Soc.Am.72, 327-330, 1982.
[84] PROVENCHER,S.W.,VOGEL,R.H.: Regularization Techniques for
     Inverse Problems in Molecular Biology.
     In [27],p.304-319.
[85] RAEDE,J.B.: Eigenvalues of positive definite kernels.
     SIAM J.Math.Anal.14, 152-157, 1983.
[86] RAEDE,J.B.: On the sharpness of Weyl's estimate for eigen-
     values of smooth kernels. SIAM J.Math.Anal.16,548-550,1985.
[87] RICHTER,G.R.: Numerical solution of integral equations of
     the first kind with nonsmooth kernels. SIAM J.Numer.Anal.
     15, 511-522, 1978.
[88] RICHTER,G.R.: An inverse problem for the steady state
     diffusion equation. SIAM J.Appl.Math.41, 210-221, 1981.
[89] SABATIER,C.P.: Positivity constraints in linear inverse
     problems - I.General theory. Geophys.J.R.astr.Soc.48,
     415-441, 1977.
[90] SABATIER,P.C.: Applied inverse problems. Lect.Notes in
     Physics, Vol.85, Springer-Verlag, 1978.
[91] SAUTTER,W.: Fehleranalyse für die Gaußelimination zur Be-
     rechnung der Lösung minimaler Länge. Numer.Math.30, 165-
     184, 1978.
[92] SANTORO,R.J.,SEMERJIAN,H.G,EMMERMAN,P.J.,GOULARD,R.:
     Optical tomography for flow field diagnostics. Int.J.
     Heat Mass Transfer 24, 1139-1150, 1981.
[93] SCHABACK,R.: Suboptimal exponential approximations. SIAM
     J.Numer.Anal.16, 1007-1018, 1979.
[94] SCHARLACH,R.: Optimal recovery by linear functionals.
     J.of Approximation Theory 44, 167-172, 1985.
[95] SHEPP,L.A.: Computerized Tomography and Nuclear Magnetic
     Resonance. J.Computer Assisted Tomography 4, 94-107,1980.
[96] SLUIS,v.d.A.,VELTKAMP,G.W.: Restoring rank and consist-
     ency by orthogonal projection. Linear Algebra and its
     Applications 28, 257-278, 1979.
[97] SMITHIES,F.: Integral equations. Cambridge Univ.Press,
     London, 1958.
[98] THOWSEN,A: Identifiability of dynamic systems. Int.J.
     Systems Sci. 9, 813-825, 1978.
[99] UTRERAS,F.D.: Optimal smoothing of noisy data using
     spline functions. SIAM J.Sci.Stat.Comput.2, 349-362,1981.
[100] VERETENNIKOV,A.Yu.,KRASNOSEL'SKII,M.A.: Regularizing
     stopping rules under conditions of random errors. Sov.
     Math.Dokl.27, 90-94, 1983.
[101] WEDIN,P.Å.: Perturbation theory for pseudo-inverses.
     BIT 13, 217-232, 1973.
[102] WEIDMANN,J.: Linear operators in Hilbert spaces.Springer-
     Verlag, New York, 1980.
[103] WENDLAND,W.: On Galerkin collocation methods for integral
     equations of elliptic boundary value problems. ISNM 53,
     244-275, 1979.

[104] WEYL,H.: Das Asymptotische Verteilungsgesetz der Eigen-
      werte linearer partieller Differentialgleichungen. Math.
      Ann. 71, 441-479, 1912
[105] WIKSWO,J.P.,MALMIVUO,J.A.V.,BARRY,W.H.,LEIFER,M.C.,
      FAIRBANK,W.M.: The theory and application of magnetocar-
      diography. Adv.in Cardiovascular Physics 2, 1-67,1979.
[106] WING,G.M.: Condition numbers of matrices arising from
      the numerical solution of linear integral equations of
      the first kind.J.of Integral Equations, to appear.
[107] YOULA,D.C.: Generalized image restoration by the method
      of alternating orthogonal projections. IEEE Trans.
      CAS-25, 694-702, 1978.

NOTATIONS

□	the end of a proof
*	the end of a remark, example, illustration
$\|\cdot\|$	any norm in $\mathbb{R}^n$
Θ	zero vector in a linear space
$B_r(x)$	ball of radius r with center x
span(....)	linear hull
int(M)	topological interior of a subset M
$\overline{M}$	topological closure of a subset M
dim(M)	dimension of a space M
A^t	the transpose of a matrix A
rank(A)	rank of a matrix
tr(A)	trace of a matrix
A^+	pseudo-inverse of a matrix
I	identity operator (matrix)
B(X,Y)	space of linear bounded operators from X into Y
B(X)	B(X) := B(X,X)
D(A)	domain of definition of a linear operator
N(A)	null-space of a linear operator
R(A)	range of a linear operator
$\oplus$	orthogonal sum
$M^\perp$	orthogonal complement of a space M
P_M	orthogonal projector onto the set M in a Hilbert space
dist(x,M)	distance of an element x from a set M
$o(\cdot),O(\cdot)$	Landau symbols
C(M;N)	the space of continuous functions from M into N
$L_p(M)$	the space of p-integrable functions from M into $\mathbb{R}$
$L_p(M;N)$	the space of measurable functions from M into N with $\|f\|^p \in L_1(M)$

SUBJECT INDEX

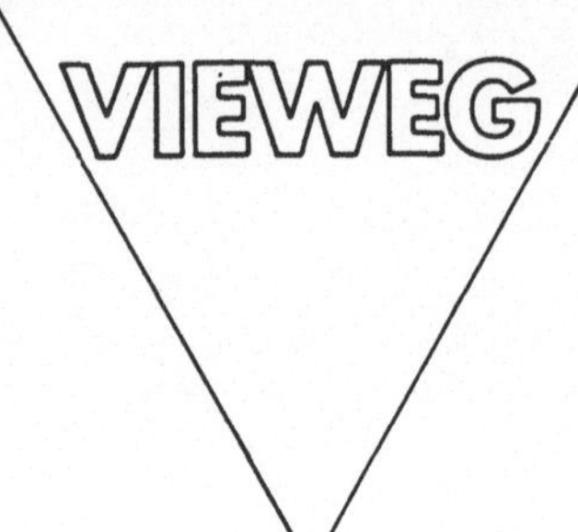

Jochen Werner

**Optimization.
Theory and Applications**

1984. VIII, 233 S. 16,2 x 22,9 cm. (Advanced Lectures in Mathematics.) Kart.

<u>Contents:</u> Introduction, Examples, Survey – Linear Programming – Convexity in Linear and normed linear Spaces – Convex optimization problems – Necessary optimality conditions – Existence theorems for solutions of optimization problems – Bibliography – Symbol Index – Subject Index.

This book is intended to give the reader an introduction to the foundations and an impression of the applications of optimization theory. It particularly emphasizes the duality theory of convex programming and necessary optimality conditions for nonlinear optimization problems. Abstract theorems are made more concrete by numerous examples from e.g. approximation theory, calculus of variations and optimal control theory. With these examples and by emphasizing the geometric background we strive to give the reader a not merely formal understanding of the subject.